THE ARLINGTON

DICTIONARY OF

ELECTRONICS

ARLINGTON BOOKS LONDON

THE ARLINGTON DICTIONARY OF ELECTRONICS

First published 1971 by
Arlington Books (Publishers) Ltd
38 Bury Street St James's
London S.W.1

© 1969 by Funk & Wagnalls,
A division of Reader's Digest Books, Inc.

Made and printed in England by
The Garden City Press Ltd
London and Letchworth

ISBN 0 85140 202 X

THE ARLINGTON

DICTIONARY OF

ELECTRONICS

Contents

EDITORIAL STAFF

EDITOR IN CHIEF
Harold A. Rodgers, Jr.

ASSOCIATE EDITORS
Michael Knibbs
Elizabeth T. Salter

EDITORIAL ASSISTANTS
Jean Kahn
Elliot Linzer

PRODUCTION EDITOR
Barbara A. Tieger

ILLUSTRATIONS
Mark Portman

TECHNICAL CONSULTANT
Norman Gleicher

Introduction
by Harold A. Rodgers, Jr.

In the technological explosion of the mid-twentieth century electronics has been one of the most active fields, particularly since solid-state devices (transistors, integrated circuits, and the like) have come into their own. At present it is possible and even likely that fields as diverse as accounting, medicine, fine arts, social sciences, and entertainment, as well as industry and the physical and life sciences, will depend increasingly heavily on electronic instrumentation. The range of application of the electronic digital computer alone appears virtually infinite.

As a result of this the terminology of electronics is no longer, if indeed it ever was, the exclusive sphere of specialists in the field. The user of electronic instrumentation has an ever increasing need to understand seemingly obscure or esoteric terms, as does the serious experimenter or hobbyist. This book is written in the hope of providing assistance to any of the divers individuals who find themselves in this position.

We have attempted in preparing this work to take a middle line between 'popular,' oversimplified definitions and definitions which, while complete, are very highly technical. For instance, while few definitions require a knowledge of mathematics for their understanding, mathematical descriptions are freely provided at places where they were deemed to be helpful. So far as possible we have provided definitions of the maximum generality. We believe this to be of particular importance, for electronics does not exist in a vacuum by itself but is a field of applications. Thus we have tried to relate electronics to basic physical facts; the *watt*, for example, is called a unit of *power*, not of *electrical power*. In addition, we have included a number of definitions that touch on particular applications in fields such as data processing, acoustics, and medicine. While some of the concepts dealt with (and, therefore, their definitions) are of considerable complexity, we have tried in each case to arrange the material in an entry in order of increasing difficulty. This should allow any reader to start at the beginning, read as far as he can, and obtain some understanding of the term; on the other hand, it does not penalize the reader who has more technical sophistication.

The attempt has been made to organize the book as compactly as possible. It is our belief that a serious reader consults a reference work not when he comes across a term that is obvious upon a moment's reflection, but only when he has a basic difficulty of comprehension. Thus the usual runs of phrasal entries beginning with words like *automatic* or *electrical* have been greatly shortened. This is not to say that phrasal entries have been excluded; they are present whenever, in the opinion of the editors, their meanings are not readily predictable.

It has been said that a dictionary is obsolete at its publication date. This is perhaps especially true of a dictionary of electronics. By relating our entries wherever possible to the root sciences from which electronics is derived, we hope, however, to retard the obsolescence of the book. Throughout we have been guided by the spirit of assistance to our readers, and while we make no claim that this dictionary is a definitive or final authority, we sincerely hope that our work will be of more than a little usefulness.

A

a *abbr.* Ampere(s).

A *symbol* **1 (a)** Linear acceleration: a vector. **2** (A) Argon. **3** (A or Å) ANGSTROM UNIT. **4** (A) Area. **5** (A) Amplification. **6** (A) AMPLITUDE. **7 (A)** Magnetic vector potential. **8** (a) A general constant.

a- *prefix* Not; without: *asymmetric, asynchronous.*

A.A.E.E. *abbr.* American Association of Electrical Engineers.

abac *n.* A NOMOGRAM.

abampere *n.* In the centimeter-gram-second system, the measure of electric current, defined as equivalent to 10 AMPERES.

A battery The battery that supplies power to heat the filament of an electron tube.

abcoulomb *n.* In the centimeter-gram-second system, the measure of electric charge, defined as equal to 10 COULOMBS.

aberration *n.* The failure of a lens to bring all the light that enters it to the same focus; also, in any system that is an analog of a lens the failure to produce a single focus.

abfarad *n.* In the centimeter-gram-second system, the measure of capacitance, defined as equal to 10^9 FARADS.

abhenry *n.* In the centimeter-gram-second system, the measure of inductance, defined as equal to one millionth of a HENRY.

abnormal glow In a glow tube, a glow that completely surrounds the cathode, resulting when the flow of current reaches a certain magnitude. Should the current increase beyond this point the current density increases correspondingly as does the voltage drop across the tube.

abohm *n.* In the centimeter-gram-second system, the measure of resistance, defined as equal to one millionth of an OHM.

abscissa *n.* In a system of rectangular coordinates, the horizontal coordinate.

absolute address In a digital computer, the actual machine address at which a particular item is stored.

absolute temperature Temperature as reckoned on the KELVIN SCALE.

absolute value **1** Of a REAL number: the value a number has if its sign is disregarded, that is, absolute value of p is given by $|p| = + p, p \geq 0$; $|p| = - p, p \leq 0$.
2 Of a COMPLEX NUMBER: the magnitude of a line connecting the origin and the complex number when it is plotted with its real part the abscissa and its imaginary part the ordinate, that is, if

$$z = x + iy, \ |z| = \sqrt{x^2 + y^2}.$$

absolute zero The temperature at which a body would be wholly deprived of heat; the zero of the Kelvin scale; $-273.16°$ C. or $-459.69°$ F.

absorb *v.* To take in and cause to disappear; in particular: **1** To take in (energy) in one form and dissipate in another. **2** To take in (a particle, substance, etc.) and bind firmly into the whole.—**absorption** *n.*

absorptance *n.* The ratio of the radiant flux absorbed by a body to that which strikes it.

absorption marker A device that identifies a particular frequency by creating a null due to the energy it absorbs at that frequency.

absorption wavemeter A frequency-measuring instrument in which a resonant element is tuned until its frequency coincides with that of the source under test, this point being indicated by a maximum absorption of energy from the source.

absorptivity *n.* ABSORPTANCE.

A-B test A test in which two components are compared by substituting one for the other in the same system and observing the overall change in the response of the system, usually used in comparing high-fidelity sound components.

AC, A.C., ac, a-c, a.c. *abbr.* Alternating current.

ACA *abbr.* Adjacent-channel attenuation.

ACC *abbr.* **1** Automatic chrominance control. **2** Automatic contrast control.

acceleration *n.* **1** The rate **(a)** at which velocity changes with respect to time. In this sense acceleration is a vector given by

$$\mathbf{a} = d\mathbf{v}/dt,$$

or for average acceleration

$$\mathbf{a} = (\mathbf{v}_2 - \mathbf{v}_1)/(t_2 - t_1).$$

2 The rate at which speed changes with respect to time. This sense is considered somewhat loose and is restricted to the description of things that happen at some speed but do not undergo displacement.—**accelerate** *v.*

accelerator *n.* A person or thing that accelerates; in particular, a device that raises charged atomic particles to high velocities (and therefore high energies) so that the manner in which they interact with atomic nuclei can be studied.

accelerometer *n.* An instrument that can be coupled to a movable system in order to sense accelerations that the system undergoes.

acceptor circuit A circuit that has been adjusted to present a maximum admittance at some particular frequency.

acceptor impurity An element whose atoms attract electrons from the crystal lattice of a semiconductor, giving rise to HOLES. These holes act as current carriers and turn the crystal into P-TYPE SEMICONDUCTOR material.

access time In a digital computer: **1** The time lapse between the instant when the arithmetic unit interrogates storage and the instant when the required information is delivered. **2** The time lapse between the instant when the arithmetic unit transmits data to storage and the instant when storage is completed.

AC coupling A method of coupling two circuits so that only the currents and/or voltages that change with time are passed from one to the other, often by the use of a capacitor or transformer.

accumulator *n.* In a digital computer, a REGISTER in which the results of addition and of other arithmetical and logical operations are stored.

accuracy *n.* In computer usage, freedom from errors. Compare PRECISION.

acetate *n.* Cellulose acetate, a plastic used in making disk records and as the base for magnetic recording tapes.

acorn tube A small vacuum tube for use at ultra-high frequencies. The small size and close spacing of the electrodes reduces interelectrode capacitance and transit time.

acoustic or **acoustical** *adj.* Of, concerned with, or related to sound, whether audible or not.—**acoustically** *adv.*

acoustic feedback In a sound-reproducing or amplifying system, the transfer of sound from the output transducer, (usually a loudspeaker) back to the input transducer (a microphone, phonograph pickup, etc.), causing the system to oscillate and produce a spurious output.

acoustic (or **acoustical**) **impedance** A quantity analogous to electrical IMPEDANCE, specifically, the complex ratio of (SOUND PRESSURE) to VOLUME VELOCITY. Acoustic impedance has a real part (**acoustic resistance**) and an imaginary part (**acoustic reactance**). The **acoustical ohm** is defined as that value of acoustic impedance, resistance, or reactance through which a sound pressure of 1 microbar produces a volume velocity of 1 cc per second. It is convenient to define **acoustic compliance** (C_a) which is analogous to capacitance, so that its associated reactance (X_c) is given by

$$X_c = 1/\omega C_a.$$

This gives it the dimensions (length)^5/force. Similarly **acoustic mass** or **inertance** (M) which is analogous to inductance has the dimensions mass/(length)^4. **Acoustic stiffness** is the reciprocal of acoustic compliance. See DYNAMIC ANALOGY.

acoustic labyrinth A loudspeaker enclosure in which the rear of the loudspeaker is coupled to a tube which at the resonant frequency of the loudspeaker is one quarter of a wavelength long. The tube, folded upon itself in order to save space, gives the appearance of a labyrinth.

acoustic lens A diffraction grating used in conjunction with a high-frequency loudspeaker (tweeter) in order to improve the dispersion of sound.

acoustics *n.* 1 The branch of physics that is concerned with the study of sound. 2 The sound-transmission qualities of an enclosed space, as a room, auditorium, etc.

acoustic suspension A loudspeaker system in which the moving cone is held by an over-compliant suspension, the stiffness required for proper operation being supplied by air that is trapped behind the cone in a sealed enclosure. Such systems, while relatively inefficient, permit good bass reproduction in a unit of moderate size.

acquire *v.* To locate and begin to track (an object) by means of radar.—**acquisition** *n.*

ACSR *abbr.* Aluminum conductor, steel-reinforced.

actinic *adj.* Designating electromagnetic radiation able to produce chemical changes.

actinium (Ac) Element 89; Atomic wt. 227; Melting pt. 1050°C. Radioactive with a half-life of 21.7 years; emits beta rays, yields protactinium.

active *adj.* 1 Of an atom or element: **a** Radioactive. **b** Fissionable. 2 Of a component, network, etc.: Receiving power from a source other than its signal input. 3 Directly involved with the performance of a specified function; necessary even in an ideal case. 4 Available for WORK.—**activate** *v.*—**activity** *n.*

active lines In a television picture, the scanning lines that actually carry picture information, as distinguished from those occurring during retrace time.

ACW *abbr.* Alternating continuous waves.

Adcock antenna An antenna whose directional pattern resembles a figure eight, consisting of two phase-opposed vertical elements located one half wavelength or less apart.

adder *n.* A device that is able to form the sum of two or more numbers. See HALF-ADDER.

additron *n.* A vacuum tube in which a beam of electrons may be rapidly switched back and forth, used as a binary adder in some computers.

address *n.* In a digital computer, an integer or set of characters that refers to the location of a particular piece of information in STORAGE.

ADF *abbr.* Automatic direction finder.

admittance *n.* The ratio (Y) of current to voltage in an electric network; the reciprocal of IMPEDANCE.

advance CONSTANTAN.

advance ball In disk recording, a rounded support that rides on the recording surface just ahead of the cutting stylus. It is used to compensate for irregularities in the surface.

advanced license A license issuable by the FCC to ham radio operators who are capable of sending and receiving code at the rate of 13 words per minute, and are familiar with general and intermediate radio theory and practice. Its privileges include exclusive use of certain frequencies.

aeolight *n.* A glow discharge lamp having a cold cathode and whose output of light is proportional to the voltage applied to its terminals, used to make a beam of modulated light for recording motion-picture sound tracks.

aerial *n.* An antenna.

A.E.T. *abbr.* Associate in Electrical Technology.

A.F., a.f., af, Af *abbr.* Audio frequency.

AFC *abbr.* AUTOMATIC FREQUENCY CONTROL.

afterglow *n.* **1** In a gas-discharge tube, the continued emission of light after the voltage has been disconnected from its terminals. **2** In a cathode-ray tube, the continued emission of light from the phosphorescent screen after the electron beam has passed.

AGC, agc *abbr.* Automatic gain control.

age *v.* To store (an electrical component) in a specified environment, as with respect to pressure, temperature, applied voltage, etc., until its characteristics stabilize.

agonic line One of the lines on the earth's surface along which true north and magnetic north coincide.

AGS *abbr.* Automatic gain stabilization.

AH, amp-hr, a.h. *abbr.* AMPERE-HOUR(S).

alacritized switch A mercury switch in which the tendency of the mercury to adhere to the surface over which it rolls has been reduced.

alchrome An alloy of 79% iron, 16% chromium, 5% aluminum. High resistivity alloy; oxidation-resistant to about 1315°C.

alcres An alloy of 83% iron, 12% chromium, 5% aluminum. High resistivity alloy; oxidation-resistant to about 1150°C.

ALGOL *n.* An algebraic and logical programming language used by many international computer groups.

algorithm *n.* In mathematics, a definite step-by-step process for finding the solution to a problem or evaluating a function.

alignment *n.* The adjustment of various components of a system so as to produce the correct overall response, in particular: **1** The tuning of frequency-selective elements in a radio receiver. **2** Adjustment of synchronization.—**align** *v.*

allo- *combining form* Other; different; extrinsic.

allochromatic *adj.* Exhibiting photoelectric effects due to the inclusion of microscopic impurities, or as a result of exposure to various types of radiation. Compare IDIOCHROMATIC.

alloy 1 *n.* A mixture of two or more elements, at least one of which is a metal, that is homogeneous in its macrostructure. **2** *v.* To unite (a metal) with another element in an alloy.

alloy junction A semiconductor junction that is formed by alloying a small amount of a DONOR or ACCEPTOR IMPURITY with a crystal of an INTRINSIC SEMICONDUCTOR material.

alloy 1040 A soft magnetic alloy of 11% iron, 3% molybdenum, 14% copper and 72% nickel having initial and maximum permeabilities of 40,000 and 100,000 gauss/oersted, a coercive force of 0.02 oersted and a Curie point of 290°C.

all-pass filter or **all-pass network** A network designed to produce a delay (phase shift) and an attenuation that is the same at all frequencies; a lumped-parameter delay line.

all-wave antenna An antenna designed to perform acceptably over a wide band of frequencies, often the standard broadcast and short-wave frequencies. In general, considering the stringent requirements of modern telecommunications systems, it is best to avoid a term as ambiguous as this.

alnico *n.* The name of a series of alloys used for permanent magnets, containing iron, cobalt, nickel, copper and aluminum, and occasionally titanium and niobium. Alnico V has the highest BH product of any known alloy except Pt-Co with a value of 6.0×10^{-6} gauss-oersteds. Its components are 51% iron, 24% cobalt, 14% nickel, 8% aluminum and 3% copper. Coercive force 660, remanence 13,100. Alnico XII (33% Fe 18% Ni 35% Co 6% Al 8% Ti) has the highest coercive force of any alnico.

alpha (*written* A, α) The first letter of the Greek alphabet, used symbolically to represent any of various coefficients, constants etc., of which in electronics the principal ones are: **1** (α) An angle. **2** (α) Angular acceleration. **3** (α) Attenuation constant. **4** (α) Current amplification. **5** (α) ALPHA PARTICLE (which see).

alpha cutoff frequency In a transistor, the frequency at which the current amplification produced in the common-base mode (α) drops 3 decibels below the low frequency value, that is, a frequency, f where

$$\alpha(f) = (1/\sqrt{2})\,\alpha_0.$$

alphameric or **alphanumeric** *adj.* Of or consisting of alphabetic characters, numerals, punctuation, mathematical symbols, etc.

alpha particle A positively charged particle consisting of two protons and two neutrons. It is identical with a helium nucleus.

alt *abbr.* Altitude.

alternating current An electric current that periodically reverses its direction of flow.

alternation *n.* A portion of a cycle of alternating current during which the current reverses its direction of flow.

alternator *n.* A rotary machine that uses the principle of electromagnetic induction to convert mechanical energy into electrical energy in the form of an alternating current.

Aluchrom O An alloy of 65% iron, 30% chromium, 5% aluminum. High resistivity alloy; oxidation resistant to about 1315°C.

Alumel *n.* An alloy of 94% nickel, 2% aluminum, 3% manganese and 1% silicon used with CHROMEL P in high temperature thermocouples.

aluminum (Al) Element 13; Atomic wt. 26.97; Melting pt. 660°C; Boiling pt. 2057°C; Density (2057°C) 2.7 g/ml; Hardness 3; Electrical resistivity (20°C volume) 2.8×10^{-6} ohm/cm^2; Spec. heat 0.215 cal/(g)/(°C). Used in electronic equipment as an electrical conductor, for capacitor foil, tuning capacitor plates, chasis, shields, etc.

aluminum antimonide A crystalline compound, AlSb, having useful semiconductor properties, which it retains to about 500°C.

AM *abbr.* Amplitude modulation.

A/m *abbr.* Ampere(s) per meter.

amateur *n.* One who pursues an activity for pleasure rather than as a profession, in particular, a duly licensed person who operates a radio station as a hobby.

amateur extra license A license issuable by the FCC to ham radio operators who are able to send and receive code at the rate of 20 words per minute, and who are familiar with general, intermediate, and advanced radio theory and practice. Its privileges include all authorized amateur rights and exclusive right to operate on certain frequencies.

ambiguity *n.* In a control system, servomechanism, etc., the tendency to stabilize at more than one point. Compare HUNTING.

American Wire Gauge A standard system of wire diameters used in the United States. See table p. 229.

americium (Am) Element 95; Atomic wt. 243; Density (0°–20°) 11.7; Valence 3, 4, 5, or 6; Spec. heat 0.033 cal/(g)(°C). Radioactive with a half-life of 458 years; emits beta rays, yields neptunium-237.

amp *abbr.* Ampere(s)

amp *n. informal* An amplifier.

ampere *n.* The measure of electric current, defined as a flow of electric charge at the rate of one COULOMB per second. [See AMPÈRE'S LAW.]

ampere-hour A measure of electric charge applied in particular to the rating of storage batteries, defined as the charge transported by a current of one AMPERE flowing for one hour, that is, 3600 COULOMBS.

Ampère's law A VECTOR equation relating current flow and magnetic fields. The law states that if a current (I) flows through an infinitesimal distance ($d\,l$) along a line, at some point (P) a distance (r) away there is produced an infinitesimal element of a magnetic field ($d\,\mathbf{H}$) such that

$$|d\,\mathbf{H}| = I \sin\theta\, d\,l/|\mathbf{r}^2|,$$

where θ is the angle between the direction of current flow and the line joining P and $d\,l$. More compactly,

$$d\,\mathbf{H} = I d\,l \times \mathbf{r}/r^3.$$

Ampère's rule A statement that follows mathematically from Ampère's law, namely that if a current flows in a wire directly toward an observer, the magnetic flux (with respect to the observer) circulates counterclockwise around the wire.

ampere-turn A measure of magnetomotive force, especially as developed by an electric current, defined as the magnetomotive force developed by a coil of one turn through which a current of one ampere flows, that is, 1.26 GILBERTS.

amplidyne *n.* A special type of dynamo used as a high-gain magnetic power amplifier, often used in servo systems.

amplification factor In a vacuum tube, a parameter (μ) defined as the negative of the ratio of a small change in plate voltage to the small change in grid voltage associated with it while the plate current is held constant, more precisely

$$\mu = (-\partial V_p / \partial V_g)$$

i_p, constant.

amplifier *n.* Any device that allows an input signal to control an external source of power in such a way that an output signal is produced that is identical to the input signal in every way except for its greater amplitude or power. By the use of DYNAMIC ANALOGY where necessary any amplifier can be represented in "black box" form as an electrical network having one pair of input terminals and one pair of output terminals. In general, all the parameters of the amplifier can be defined in terms of the currents and voltages at the terminals given above. Amplifiers can be classified in terms of their applications, operating parameters, active components, etc. See CLASS A, CLASS B, etc.; LINEAR AMPLIFIER, MAGNETIC AMPLIFIER, PARAMETRIC AMPLIFIER, POWER AMPLIFIER, PUSH-PULL AMPLIFIER, TRANSISTOR AMPLIFIER, VACUUM-TUBE AMPLIFIER.—**amplify** *v.*

amplitude *n.* **1** The peak excursion of a periodically varying quantity away from its mean or reference position or value. **2** Of a complex number $z = x + iy$, an angle, θ, such that

$$z = r (\cos \theta + i \sin \theta), \text{ where } r = |z|.$$

amplitude modulation The variation of the amplitude of a wave in a way that corresponds to another wave. In general, the positive or negative envelope of the modulated wave (carrier wave) contains enough information to allow the modulating wave to be recovered, provided that the carrier wave has at least twice the frequency and twice the peak amplitude of the modulating signal. The modulated carrier $C(t)$ is given by

$$C(t) = A_0 [1 - k M(t)] \cos \omega_0 t,$$

where $M(t)$ is the modulation.

AMT *abbr.* Amplitude-modulated transmitter.

an- *prefix* Not; without: *anacoustic, anechoic.*

anacoustic *adj.* Without sound, designating in particular the regions on the fringes of the earth's atmosphere and beyond, where sound propagation is negligible or zero.

analog or **analogue** **1** *n.* Any of two or more systems that are mutually related in that their mathematical descriptions are essentially equivalent. For instance, in a simple mechanical system the resonant frequency (f_0) is given by

$$f_0 = 1/(2\pi \sqrt{mc}),$$

where m is MASS and c COMPLIANCE, in an electrical system

$$f_0 = 1/(2\pi \sqrt{LC}),$$

where L is INDUCTANCE and C CAPITANCE. **2** *adj.* Of, being, or relating to information in the form of variables that undergo smooth, continuous variation. This term alludes to the fact that an analog computer can accept information in continuous form while a digital computer requires that there be discrete separations, however small, between successive values of a variable. **3** *v.* To substitute a convenient analog for (a system under study) in order to simplify analysis. In general, electrical analogs are most convenient and are most widely used. See DYNAMIC ANALOGY; DUALITY.

analog computer An electronic device used to simulate the mathematical equations describing a system under study. The readout takes the form of a continuously varying voltage which is usually displayed on a strip chart recorder or oscilloscope. The basic building block of the analog computer is the OPERATIONAL AMPLIFIER which in conjunction with other electrical components form necessary functional devices such as INTEGRATORS, SUMMERS, MULTIPLIERS, and INVERTERS. Analog computers can solve, in general, many problems which are difficult or impossible by analytic means. They have a lower accuracy capability than digital machines but offer a faster solution which is continuous rather than discrete.

analyzer *n.* Any of various instruments used in electronic testing and analysis. See DISTORTION ANALYZER, SPECTRUM ANALYZER.

Anderson bridge A six-element BRIDGE used to measure inductance in terms of resistance and capacitance.

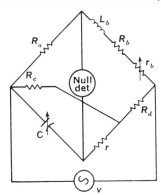

$$L_b = \left(R_d + r + \frac{rR_d}{R_c} \right) R_a C$$

$$R_b + r_b = \frac{R_a R_d}{R_c}$$

AND-gate A LOGIC CIRCUIT having two or more inputs and a single output. A signal appears at its output if and only if all of its inputs are energized simultaneously.

anechoic *adj.* Nonreflective; producing no echos.

anechoic room A room whose walls have been treated so as to make them absorb a particular kind of radiation almost completely, used for testing components of sound systems, radar systems, etc., in an environment free of reflections.

anelectrotonus *n.* The reduced sensitivity produced in a nerve or muscle in the region of contact with the anode when an electric current is passed through it. —**anelectrotonic** *adj.*

angel *n. informal* A radar reflection of short duration in the lower atmosphere corresponding to no observable object. Investigations have shown that these are in many cases caused by insects or birds.

angle modulation The modulation of a sinusoidal carrier wave by variation of its angle, that is, given a carrier, s(t), where

$$s(t) = A \cos (\omega t + \phi),$$

modulation is accomplished by varying ω and ϕ. FRE-QUENCY MODULATION and PHASE MODULATION are particular forms of angle modulation, and are related by the fact that instantaneous frequency is the derivative with respect to time of instantaneous phase.

angstrom or **anstrom unit** A unit of length equal to 10^{-8} centimeter or 3.937×10^{-9} inch.

angular acceleration The rate (α) at which angular velocity (ω) changes with respect to time, generally expressed in radians per second per second and given by

$$\alpha = d\omega/dt = d^2\phi/dt^2$$

angular distance Distance as expressed in angular measure, generally in radians. Angular distance (in radians) is equal to $2\pi l/\lambda$, where l is linear distance and λ is the wavelength.

angular frequency Frequency expressed in RADIANS per second rather than cycles per second (hertz). As there are 2π radians in each cycle, angular frequency (ω) is equal to 2π times the frequency in hertz.

angular momentum The MOMENTUM that a body has by virtue of its rotational movement. In vector notation, the angular momentum **(L)** of a particle of mass m about an origin is given by

$$\mathbf{L} = m(\mathbf{r} \times \mathbf{v}) = \mathbf{r} \times \mathbf{p},$$

where **r** is the position vector with respect to the origin, **v** is the linear velocity, and **p** the linear momentum. Alternatively $\mathbf{L} = I\omega$, where I is the MOMENT OF INERTIA of the particle (or of a system of particles) and ω is the ANGULAR VELOCITY.

angular velocity The rate (ω) at which an angle changes with respect to time, generally expressed in radians per second, that is, $\omega = d\phi/dt$, where ϕ is the angle and t time. Angular velocity usually refers to rotational movement. Its instantaneous value may be given as a vector (ω) of magnitude whose positive sense is in the direction of the axis of rotation about which the rotation appears clockwise.

anharmonic oscillator An oscillating system in which the restoring force is a nonlinear function of the displacement from equilibrium.

anion *n.* A negatively charged ion, especially of an electrolyte.

anisotropic *adj.* Having characteristics, as wave propagation constant, magnetic permeability, conductivity, etc., that vary with direction; not isotropic.—**anisotropy** *n.*

ANL *abbr.* Automatic noise limiter.

anneal *v.* To heat (a solid material) and allow it to cool, usually slowly, altering its internal structure in a way that renders it softer and less brittle.

annular *adj.* Having the form of a ring; ring-shaped.

annunciator *n.* A system in which each of several calling stations can signal a single receiving station, the source of each signal being indicated at the receiver.

anode *n.* The electrode through which current flows into a device, that is, the positive terminal of a device that uses electricity, or the negative terminal of a device that produces electricity; in British terminology, also the PLATE of an electron tube. A device such as a linear bilateral resistor, which is indifferent as to the direction of the current through it would not normally be said to have an *anode* or a *cathode.*

anode metal An alloy of 98.8% lead, 1% silver and 0.2% arsenic used in lead storage battery plates.

anodize *v.* To deposit a protective coating of oxide on (a metal) by means of an electrolytic process in which it is used as the anode.

anomalous propagation Unusual or abnormal wave propagation; in particular: **1** The propagation of UHF signals through ductlike passages formed by layers of the atmosphere, by means of an action similar to that of a waveguide. **2** Irregular variations in the strength of a sonar echo, caused by rapid variations in the propagation characteristics of the water.

anotron A glow-discharge diode having a cold cathode, often made of sodium. The anode is generally of copper.

ant. *abbr.* Antenna.

antenna *n.* A device designed to radiate an electromagnetic wave into space when excited by a radio-frequency alternating current, or, alternatively, to absorb energy from an electromagnetic wave and produce a radio-frequency alternating current. Generally, an antenna consists of a conductor or system of conductors.

antenna field gain A figure of merit that measures the effectiveness of a transmitting antenna, defined as the ratio of the field intensity in the horizontal plane produced at a distance of one mile with an input power of 1 kilowatt to 137.6 millivolts per meter.

anti- *prefix* **1** Against; opposed to: *antinoise; antifriction.* **2** Inverse of; opposite to: ANTILOGARITHM; ANTIPARTICLE.

anticathode *n.* In an x-ray tube, the target that is bombarded by a beam of electrons and which, in turn, emits x-rays.

anticoincidence circuit A NAND-gate.

antiferroelectricity *n.* A property of certain crystals whereby they undergo a phase transition from a condition of low symmetry to one of high symmetry as temperature is varied. Unlike ferroelectrics these crystals lack permanent electrical polarization.—**antiferroelectric** *adj., n.*

antiferromagnetism *n.* A property of certain metals, alloys, and salts whereby at sufficiently low temperatures the neighboring atomic magnetic moments are arranged in an antiparallel fashion rather than parallel as in the case of ferromagnetic materials.—**antiferromagnetic** *adj.*—**antiferromagnet** *n.*

antilogarithm or **antilog** *n.* The number (*n*) corresponding to a given logarithm, that is,

$$n(q) = b^q,$$

where *q* is the logarithm and *b* its base.

antimony (Sb; *Lat.* stibium) Element 51; Atomic wt. 121.76; Melting pt. 630.5°C; Boiling pt. 1380°C; Spec. grav. 6.691; Hardness 3.0-3.3; Valence 3 or 5; Spec. heat 0.049 cal/(g)/(°C); Electrical conductivity (0-20°C) 0.026×10^6 mho/cm. Used as a donor impurity in semi-conductors and, as an alloy with lead, used for storage batteries.

antinode *n.* In a system of standing waves, any of the points where the amplitude has a maximum; a loop.

antineutrino *n.* The ANTIPARTICLE corresponding to the neutrino.

antineutron *n.* The ANTIPARTICLE corresponding to the neutron.

antiparticle *n.* Either of a pair of atomic particles having identical mass, equal and opposite charge, and equal spin and magnetic moment but with the spin oppositely directed with respect to the magnetic moment. If such a pair interact they annihilate each other, their combined mass appearing as equivalent energy.

antiproton *n.* The ANTIPARTICLE corresponding to the proton.

antiresonance n. A type of resonance in which a system offers maximum IMPEDANCE at its resonant frequency.—**antiresonant** adj.

antitransmit-receive tube A gas-filled tube used as a switching element in a DUPLEXER.

APC abbr. Automatic phase control.

aperiodic adj. Nonresonant; having no oscillatory tendencies.

aperiodic damping The damping of a resonant system to a sufficient extent that the system returns to rest without overshoot when subjected to a single disturbance.

aperture n. An opening through which a flow of radiation or particles may pass, in particular: **1** An opening through which light enters the lens of a camera or other optical instrument. **2** The part of a plane perpendicular to the direction of maximum radiation through which the major part of the radiation to or from an antenna passes.

aperture mask In a color-television picture tube that has three electron guns, a thin sheet of perforated metal mounted in such a way that each electron beam can strike only the phosphor for the color with which it is associated.

apparent power See COMPLEX POWER.

Applegate diagram A plot of distance against time used to represent electron bunching in a velocity-modulated microwave tube.

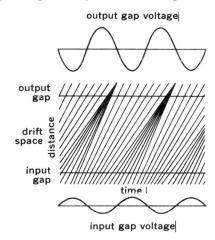

Applegate diagram for klystron illustrating bunching

Appleton layer The F layer of the ionosphere.

apple tube A color-television picture tube in which the three colors of phosphors are laid in fine vertical strips along the screen and the intensity of the electron beam is modulated as it sweeps over them so that each color is produced with appropriate brightness.

aquadag n. A conductive coating of graphite applied to the inside (and occasionally the outside) of certain tubes, cathode-ray tubes in particular. Kept at a positive potential, it often serves to absorb secondary electrons: a trade name.

arc **1** n. A segment of any curve. **2** n. An ARC DISCHARGE. **3** v. To undergo or produce an ARC DISCHARGE. **4** adj. In mathematics, inverse. Arc is used specifically to designate the inverse trigonometric and hyperbolic functions. For example, arc sin x is the angle whose sine is x. In general, arc sin x is notated as $\sin^{-1} x$, the other functions being indicated in a similar manner. When written in full the inverse functions are sometimes made one word, that is, arc sine becomes arcsine, etc.

arcback n. In a rectifier tube, heavy conduction in the reverse direction due to some internal fault, as for instance, emission of electrons from the plate due to overheating.

arc discharge A mode of electrical conduction in gases that resembles somewhat a GLOW DISCHARGE, but has a higher current density and a lower voltage drop.

arc sine or **arc sin** or **arcsin, arc tangent,** etc. See ARC, def. 4.

argon (A) Element 18; Atomic wt. 39.944; Melting pt. −189.2°C; Boiling pt. −185.7°C; Density 1.784 g/l; Valence 0; Spec. heat 0.125 cal/(g)/(°C); Used as the atmosphere of tungsten electric light bulbs, in glow lamps, and electron tubes. Also used in inert-gas-shielding arc-welding, and in electric arc cutting of non-ferrous metals.

argument n. **1** An independent variable. See FUNCTION. **2** The AMPLITUDE of a complex number.

arithmetic unit In a digital computer, the part of the hardware that performs arithmetical and logical operations.

armature *n.* **1** In an electric generator, the coil in which the electromotive force is induced. In dc generators the armature is a rotor, but in alternators it may be either a stator or a rotor. **2** In an electromechanical device such as a motor, relay, etc., the moving element.

armor *n.* A protective sheath of metal enclosing the insulation of a wire, cable, etc.

Armstrong oscillator A type of TUNED-GRID TUNED-PLATE OSCILLATOR. Feedback is delivered through the grid-plate capacitance of the tube, while the plate load is tuned slightly above the operating frequency in order that it appear inductive.

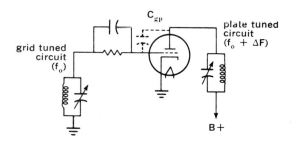

array *n.* A group of transducers driven by the same energy source in such a way as to produce a desired effect, in particular, a set of antenna elements arranged to produce a certain gain and directivity.

arsenic (As) Element 33; Atomic wt. 74.91; Melting pt. sublimes (500°C under pressure); Boiling pt. 615°C; Spec. grav. 5.73; Valence 3 or 5; Spec. heat 0.082 cal/(g)/(°C); Electrical conductivity (0−20°C) 0.029 × 10^6 mho/cm. Used as a donor impurity in semi-conductors.

articulation *n.* In a communications channel, the percentage of single-syllable nonsense words that can be understood after passing through the channel. Compare INTELLIGIBILITY.

asbestos A white or light gray fibrous mineral that has excellent insulating properties and resistance to heat.

ASC *abbr.* Automatic sensitivity control.

A-scan A cathode-ray tube display in which the horizontal deflection represents distance and the vertical deflection represents the magnitude of a reflected pulse, used in radar, sonar, and related systems.

aspect ratio In television, the width to height ratio of the frame. In the United States this has been standardized at 4:3.

asperity *n.* A microscopic projection from a surface. Each asperity on an electrode surface creates through its geometry an intensification of the electric field at its tip.

aspheric *adj.* Deviating from a spherical contour, as a lens or reflector, generally to avoid aberration.

ASRA Automatic Stereophonic Recording Amplifier.

assembler *n.* A computer program that translates a source program written in a symbolic programming language into a machine-language object program while relieving the programmer of most housekeeping tasks.

astable *adj.* **1** Not stable. **2** Having no stable states, in particular, designating a circuit that spontaneously undergoes transitions between its various states.

astatic *adj.* **1** Having characteristics that do not vary with direction. **2** Being in neutral equilibrium.

astatine (At) Element 85; Atomic wt. 211; Melting pt. 302°C; Valence 1, 3, 5, or 7. Radioactive element with isotope having half-life of 7.5 hours; decays partly by K-electron capture and partly by alpha-particle emission.

astrionics *n.* The part of electronics that is concerned with astronautical applications.

astrocompass *n.* An instrument that determines direction, using the stars as a reference.

A-supply In a device using electron tubes, the source of the electricity that is used to heat the filaments of the tubes.

asymptote *n.* A line representing the limiting case of the tangent to a given curve as one of the variables defining the points of the curve undergoes an unbounded increase, that is, given a curve defined by $f(x,y) = 0$

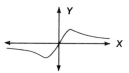

The *x*-axis is the asymptote to the curve shown.

having a set of tangents in the form $y = mx + b$, the line $y = m_0 x + b_0$ is an asymptote if and only if lim $m = m_0$ and lim $b = b_0$ as the x or y coordinate of the point of tangency increases without bound.

AT, At *abbr.*   Ampere-turn(s).

A.T.C. *abbr.*   Antenna tuning capacitor.

AT-cut crystal   A slab of quartz cut at an angle of about 35.5° with the optical axis (Z-axis) of the mother crystal. It has a temperature coefficient of virtually zero at some temperature that depends on the exact angle of the cut.

A.T.I. *abbr.*   Antenna tuning inductance.

At/m *abbr.*   Ampere-turns per meter.

atmosphere *n.*   **1** The envelope of gases surrounding a planet, star, etc. **2** A unit of pressure equal to that of the earth's atmosphere at sea level and at a temperature of 0° C., that is, 760 mm of mercury or 14.69 lb per sq. in.

atmospherics *n. pl.*   Any of various forms of radio-frequency noise resulting from natural disturbances in the atmosphere.

atom *n.*   The basic unit of a chemical element, consisting of a positively charged nucleus surrounded by a number of electrons sufficient to counterbalance the charge of the nucleus. The identity of an element, in a chemical sense, depends on the number of positive charges in the nucleus of its atom. The nucleus also contains particles that contribute mass but no charge. The stability of a nucleus depends on its ratio of charge to mass. See ISOTOPE, NUCLEUS.

atomic mass unit   A measure of atomic mass, defined as equal to 1/12 the mass of a carbon atom of mass 12. Abbr. *amu*

atomic number   A number that represents the unit positive charges (protons) in the atomic nucleus of each element and corresponds to the number of extra-nuclear electrons.

atomic weight   Since 1961, the weight of an atom of an element relative to that of an atom of carbon, taken as 12.01115.

ATR *abbr.*   Anti-transmit-receive.

ATR tube   ANTI-TRANSMIT-RECEIVE TUBE.

attack *n.*   **1** The beginning and approach to a steady-state response on the part of an electrical signal (an audio signal in particular), musical sound, etc. **2** The action of a control system in response to a sudden error condition.—**attack** *v.*

attack time   **1** The time required for a system to complete a specified fraction (generally $1 - 1/e$ or about 63 percent), of its approach to a steady-state response after a sudden disturbance. **2** The time from the initiation of a signal to the point where it reaches $1 - 1/e$ of its final amplitude. Compare DECAY TIME, FALL TIME, RISE TIME, TIME CONSTANT.

attenuate *v.*   To reduce in amplitude, intensity, etc.; make or become weaker.—**attenuation** *n.*

attenuation constant   The real part of the PROPAGATION CONSTANT, that is, a number (α) expressing the loss (either in nepers or decibels) per unit distance of a transmission system.

attenuator *n.*   A device designed to lower the amplitude of a signal by a desired amount while leaving all of its other characteristics unchanged.

atto- *combining form.*   One quintillionth (10^{-18}) of a specific quantity or dimension.

At/Wb *abbr.*   Ampere-turns per weber.

A.U., a.u., AU *abbr.*   Angstrom (unit.)

audio 1 *adj.*   Of or having to do with sound as related to the sense of hearing, and as related to electrical and electromechanical systems intended to transmit or reproduce sound. **2** *n.* An audio signal.

audio- *combining form*   **1** Of or having to do with hearing. **2** Of or having to do with sound reproduction.

audio frequency   A wave frequency corresponding to that of a sound wave within the range of human hearing. This is usually considered to be the range from 20 hz to 20 khz, although, in fact, few persons can hear this full range at normal levels of intensity.

audiometer *n.*   An instrument used to measure the acuteness of hearing, generally consisting of a source of speech signals or a variable frequency sine-wave oscillator, whose output level can be closely controlled.

audion *n.*   An early form of vacuum triode developed by Dr. Lee de Forest.

audiophile *n.*   A person who has a strong interest in high-fidelity music reproduction systems.

Auger effect   The transition of an atom from an excited state to a lower state with the emission of an electron but without radiation.

auto- *combining form* To, of, or by the thing itself: *autocombustion; autoinduction.*

autoalarm *n.* A radio receiver tuned to the international distress frequency at 500 khz, arranged so as to activate an alarm when it receives a signal.

autocondensation *n.* A method of introducing high-frequency alternating-currents into living tissue for therapeutic purposes. The patient is connected as one plate of a capacitor to which the current is applied.

autoconduction *n.* A method of introducing high-frequency alternating-currents into living tissue for therapeutic purposes. The patient is placed inside a coil and acts essentially as the secondary of a transformer.

autodyne converter A heterodyne frequency in which the local oscillator, mixer, and first intermediate-frequency amplifier are combined into a single stage containing but one active element (vacuum tube, transistor, etc.).

autoformer An AUTOTRANSFORMER.

automatic *adj.* Self-acting and self-regulating; operating without human intervention; often implying the presence of a feedback control system.

automatic frequency control A system that produces an error voltage in proportion to the amount by which an oscillator drifts away from its correct frequency, the error voltage acting to reverse the drift.

autoelectronic effect The emission of electron from a cold cathode under the influence of a sufficiently strong electric field.

automatic gain control A feedback control that reduces the gain of a system in response to a rise in the output level, thus serving to keep the output level substantially constant.

automation *n.* The development of machines and processes that (except for original programming) operate with a minimum of human control or intervention.

automaton *n.* A device that operates according to a series of programmed instructions. *pl.* **automata.**

autopilot *n.* A system used to control the course, speed, attitude, etc., of an aircraft, ship, spacecraft, or other vehicle automatically.

autoradiography *n.* The production of an image of a

radioactive source by allowing the radiation it produces to fog a photographic plate.

Autosyn *n.* A type of synchro: a trade name.

autotransductor *n.* A magnetic amplifier in which the power windings serve also as control windings.

auxochrome *n.* A chromophore that affects strongly the wavelength at which absorption occurs.

av *abbr.* Average.

availibility *n.* The fraction (expressed by a percentage) of a specified period of time during which a piece of equipment operates correctly.

a.v.c., AVC, A.V.C., avc *abbr.* **1** Automatic volume control. **2** Automatic volume compression.

a.v.e., AVE *abbr.* Automatic volume expansion.

avionics *n.* The branch of electronics that is concerned with aviation applications.

avalanche 1 *n.* The production of a heavy current in a hitherto nonconducting medium, as the electric field becomes strong enough to ionize a few atoms. These ions and free electrons then act as current carriers which are accelerated until they produce more ions by collision, the process continuing in a rapid chain reaction. **2** *adj.* Occurring because of an avalanche. **3** *adj.* Designed to operate in avalanche breakdown for at least part of the time.

avalanche breakdown 1 A nondestructive breakdown of a reverse biased p–n junction due to an avalanche. **2** A luminous discharge in a gas tube due to an avalanche.

average *n.* A number (A) that is a function of a set of numbers (x_i) and that is considered typical or representative of the set. In general

$$A(x_i) = (\sum_{i=1}^{n} q_i x_i^y / \sum_{i=1}^{n} q_i)^{1/y},$$

where q_i represents the weight given to each x_i, n is the number of x_i and y is arbitrary. If each of the q_i equals 1 and y equals 1 this average is called the ARITHMETIC MEAN; if q_i are all 1 and y equals -1 this average is called the HARMONIC MEAN; with q_i again equal to 1 and y equal to 2 this average is called the ROOT MEAN SQUARE. This formulation may be extended to cover any continuous function, $f(x)$, defined, together with its weighting function, $q(x)$, in the interval $a \leq x \leq b$. In this case

$$A[f(x)] = [\int_a^b q(x)\, f(x)^y \cdot dx / \int_a^b q(x)dx]^{1/y}.$$

See GEOMETRIC MEAN, MEAN.

average power   See COMPLEX POWER.

Avogadro's number   The actual number of molecules in one gram-molecule or of atoms in one gram-atom of an element or any pure substance. The number is 6.023×10^{23}.

AWG abbr.   AMERICAN WIRE GAUGE.

axial adj.   **1** Of, having to do with, or forming an axis. **2** Situated on or along an axis, particularly the major axis.

axis n.   **1** A line passing through the center of a body, in particular, a line with respect to which the body is symmetrical. **2** On a graph, a line along which or parallel to which the value of a variable is plotted.

Ayrton shunt   A resistive network connected in parallel with a galvanometer to provide optimum damping and variable sensitivity.

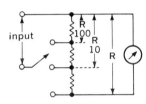

azimuth n.   Angular distance in the horizontal plane, usually as measured clockwise from north.

B

B abbr.   Break.

B symbol   **1** (B) Boron. **2** (B or b) Susceptance. **3 (B)** Magnetic flux density: a vector. **4** (b) Base (of a transistor).

B+ symbol   B-PLUS.

B− symbol   B-MINUS.

babble n.   In a communications channel, interference in the form of crosstalk from many other channels.

back diode   A BACKWARD DIODE.

backfire n.   ARCBACK.

background n.   Any of the residual random effects that a physical phenomenon must override in order to be measured, for instance, a radio telescope must discriminate between radiation from a distant star and the background of galactic noise.

backheating n.   In a magnetron, heating of the cathode as a result of electrons being returned to it at high velocities.

backlash n.   **1** A reaction or jarring recoil, as of the parts of a machine when poorly fitted or subjected to sudden strain. **2** The amount of loose play in the parts subject to such reaction or recoil.

back loading   The loading of a loudspeaker or other doublet radiator on the side that is nominally the back.

back porch   In a composite video signal, the PEDESTAL that follows the horizontal sync pulse. For illustration see COMPOSITE VIDEO SIGNAL.

backswing n.   The amplitude of the overshoot past the reference position at the trailing edge of a pulse, usually expressed as a percentage of the pulse amplitude. For illustration see PULSE.

back-to-back adj., adv.   Allowing a pair of similar unilateral components to control an alternating current, one controlling the positive alternation, the other the negative.

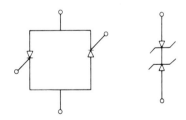

Back to back connections using (left) SCR's and (right) Zener diodes.

backup n.   An item or component that is intended to take over the function of another device in case of failure.

backward diode   A semiconductor diode that is designed to conduct heavily with a reverse bias applied.

Its current-voltage characteristics are similar to those of a TUNNEL DIODE, except that the PEAK POINT current is kept to about 50 microamperes or less.

backward-wave tube A TRAVELING-WAVE TUBE in which the phase velocity of the wave is opposite to the energy velocity of the electron beam. The positive feedback inherent in the system allows the tube to be operated as an oscillator or a narrow band regenerative amplifier at a frequency that is a function of the beam voltage.

baffle *n.* A reflective shield or partition used to guide some radiation or flux along a desired path. Also BAFFLE PLATE.

bakelite or **Bakelite** *n.* **1** Any of several phenolic plastics having good electrical insulating properties, used as materials for various sockets, circuit boards, etc.: a trade name. **2** Any of various other plastics protected under this trade name.

balanced *adj.* Indicating a network, transmission line, signal, etc., in which all components are symmetrical with respect to a common reference point, usually ground.

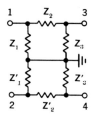

Balanced Network

balanced modulator A push-pull amplitude modulator that completely suppresses the carrier and delivers an output consisting of only the two SIDEBANDS.

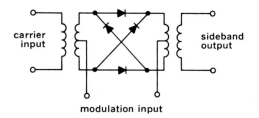

carrier input

sideband output

modulation input

balancing network A network added to a circuit of two branches to adjust it so that a voltage applied in one branch produces no current in the other.

Balco *n.* An alloy of 30% iron and 70% nickel having a high temperature coefficient of resistivity and a permeability varying linearly with temperature. It is used in electrical instruments to compensate for ambient temperature variations.

ballast *n.* An impedance connected in series with a device having a NEGATIVE RESISTANCE characteristic in order to limit the current to a non-destructive level, in particular, a resistor that is non-linear in the sense that as the current through it increases its resistance increases and tends to offset the change in current.

ballistic galvanometer A galvanometer constructed in such a way that its peak deflection (ϕ) is proportional to the charge (Q) carried by a short pulse of current. In effect the meter integrates the pulse, that is,

$$\phi \approx K \int_{t_0}^{t} I(t)dt = Q,$$

where I is current and t time, provided that $t - t_0$ is small.

balun *n.* In general, a device used to match a balanced transmission line to an unbalanced transmission, often accomplishing impedance matching as well.

banana jack A jack designed to fit a banana plug.

banana plug A single conductor electrical plug with a spring metal tip shaped somewhat like a banana.

band *n.* **1** A range of wave frequencies lying between two specified limits, in particular, a range of radio-wave frequencies reserved for a particular use. **2** A range of electron energy levels.

band center A frequency designated as the nominal center of a band, defined as the geometric mean of the extreme frequencies of the band.

bandpass filter A FILTER that in the ideal case passes all frequencies between two limits with uniform zero attenuation, while suppressing completely all frequencies that lie outside of those limits. In practice this ideal response can only be approximated.

band-rejection (or **band-elimination**) **filter** A FILTER that ideally passes all frequencies up to a certain limit, suppresses all frequencies up to another limit, and passes all frequencies thereafter.

bandspread *n.* A device, often consisting of an auxiliary tuning capacitor in parallel with the main one, that allows a radio receiver to be tuned very precisely over a narrow frequency range which is set by the main tuning control.

bandswitch *n.* A switch to select any one of the frequency bands in which a receiver, transmitter, or other device, is operative.

bandwidth *n.* **1** The numerical difference between the limits of a band of frequencies. **2** The highest sinusoidal frequency present in a complex signal.

bang-bang control system A feedback control system in which the error-correcting element has two discrete states between which it alternates as necessary.

bank *n.* A set of similar pieces of equipment connected together and arranged in a line, as transformers, lamps, etc.

bank winding A method of interwinding several layers of a coil in order to reduce the distributed capacitance. The method depends on keeping turns that tend to be at very different electrical potentials as far apart as possible.

bantam tube An electron tube having a standard octal base but an envelope much reduced in size.

bar *n.* In the cgs system, the measure of pressure, defined as equal to 1 million dynes per square centimeter.

bar generator A device that produces a train of equally spaced pulses that are synchronized in such a way as to create a series of bars when displayed on a television screen, used in making adjustments of horizontal and vertical width and linearity. For color television adjustments, each bar is made to have a different hue.

barium (Ba) Element 56; Atomic wt. 137.36; Melting pt. 850°C; Boiling pt. 1140°C; Spec. grav. (20°C) 3.5; Valence 2; Electron configuration 2 8 18 18 8 2; Ionic radius 1.35Å; Latent heat of vaporization (1140°C) 374 kilojoules/g-atom; Spec. heat 0.068 cal/(g)/(°C); Electrical Conductivity (0–20°C) 0.016 × 10⁶ mho/cm. Used in fuel cells and in cathode coatings of electron tubes.

Barkhausen criterion A statement of the conditions necessary for the operation of a feedback oscillator, namely that the gain of the feedback loop must be 1 and the phase shift must be zero at at least one frequency.

Barkhausen effect In iron and other ferromagnetic materials, the discontinuous change of magnetization that results when a continuously changing magnetizing force is applied.

Barkhausen-Kurz oscillator An oscillator in which the plate of a triode is made somewhat negative with respect to the cathode and the grid is kept positive. Electrons overshoot the grid several times before being collected, giving rise to oscillations which depend mainly on the transit time of the electrons. This oscillator is capable of operating at microwave frequencies.

Barkhausen oscillation A form of parasitic oscillation that occurs in vacuum tubes when the plate voltage drops below that of another electrode, notably the screen grid. Under these conditions an action similar to that of a Barkhausen-Kurz oscillator takes place. This difficulty is common in the horizontal output stages of television receivers.

barn *n.* A measure of CROSS SECTION for nuclear reactions, defined as equal to 10^{-24} square centimeter.

Barnett effect The slight magnetization produced in a long iron cylinder when it is rotated rapidly about its longitudinal axis. The magnetization is proportional to the angular speed, and is attributed to the effect of the rotation on the electron systems.

bar pattern The display produced on a television screen by the use of a BAR GENERATOR.

barretter *n.* A resistor, often metallic, having a positive temperature coefficient of resistivity. The resistive element is often enclosed in a gas-filled envelope. A barretter tends to pass a constant current despite variations in voltage. See BALLAST.

barrier *n.* See DEPLETION LAYER, POTENTIAL BARRIER.

base *n.* **1** In a bipolar transistor: **a** The region through which minority carriers diffuse in passing from the emitter to the collector. As some of the minority carriers recombine with the majority carriers, a flow of current is required to maintain the forward bias on the emitter-base junction. This allows the base current to control the emitter and collector currents as well. **b** The terminal connected to the base region, analogous to the grid of a vacuum tube. **2** In a UNIJUNCTION TRANSISTOR, either of the terminals at opposite ends of the n-type silicon bar. **3** A number on which a system of numbers is founded, that is, a number b such that any integer is of the form $d_0 + d_1b + d_2b^2 \ldots + d_nb^n$, where $d_1, d_2, \ldots d_n$ are non-negative integers smaller than b.

baseband *n.* The band of frequencies used to modulate a carrier.

base line On an oscilloscope, the line that the sweep produces on the cathode-ray tube in the absence of an input signal.

base spreading resistance In a transistor, the ohmic resistance of the base region, a parasitic resistance which tends to degrade the performance of the transistor.

base station In radio communications, a fixed station that communicates with one or more mobile stations.

base transport factor The degree (β^*) to which the base region of a transistor passes current carriers onto the collector region without recombinations. This parameter contributes to the gain (α) of the transistor in an important way.

bass *n.* Sounds in the low audio-frequency range.

bass reflex A type of loudspeaker enclosure in which a cavity tuned to the resonant frequency of the speaker is coupled to its rear side. The cavity damps the loudspeaker resonance and at low frequencies reverses the phase of the backward radiation and uses it to reinforce the forward radiation.

bathochrome *n.* A chromophore that lowers the frequency at which radiation is absorbed.

bathtub capacitor A type of paper capacitor enclosed in a metal casing having rounded corners.

battery grid metal An alloy 88 to 92% lead, 8 to 12% antimony and 0.25% tin used in the lead-acid storage battery in the grids supporting the active material composing the plates.

baud *n.* The measure of telegraph signaling speed, defined as a rate of one pulse (of the shortest duration used in the system) per second.

Baudot code The standard five-bit teletypewriter code.

In practice the five pulses (bits) representing any character are always preceded by a start pulse, which is always a spacing pulse (0), and are followed by a stop pulse, which is always a marking pulse (1).

bay *n.* **1** A section of an antenna array. **2** A vertical rack or housing for electronic equipment.

bayonet base A base for a lamp, tube, etc., which has two pins or prongs on opposite sides for engaging in the slots of a bayonet socket.

bayonet socket A socket designed with deep slots for engaging the pins or prongs of a bayonet base.

Bc, bc *abbr.* Broadcast.

BCB *abbr.* Broadcast band.

BCI *abbr.* Broadcast interference.

BCL *abbr.* Broadcast listener.

BD *abbr.* Bus driver.

beam *n.* **1** A flux of particles or radiation that has a relatively small cross section perpendicular to its direction of travel. **2** *v.* To emit or confine in a beam.

beam power tube A vacuum tube whose power-handling characteristics are enhanced by the addition of a suppressor grid which holds secondary emission from the plate to a minimum. In addition the control and screen grids are aligned, leaving the screen grid in the ''shadow'' of the control grid and keeping current at a minimum. Thus plate current is nearly independent of plate voltage or screen voltage.

beam-switching tube A vacuum tube containing a central cathode and several, often 10, anode arrays, each containing a collector, a beam-forming electrode, and a switching grid. Under the influence of the potential difference between the cathode and the anodes and an additional magnetic field, supplied either internally or externally, electrons flow in a beam to just one anode, the grids acting to switch the beam from anode to anode.

beat frequency Either the sum or difference frequency produced when two frequencies are heterodyned.

beat-frequency oscillator An oscillator whose output is heterodyned with another signal to produce a beat frequency.

beats *n.* A series of periodic variations of amplitude resulting from the interference of two wave trains of different frequencies.

bedspring antenna   A broadside array equipped with a flat reflector.

B.E.E. *abbr.*   Bachelor of Electrical Engineering.

bel *n.*   The ratio of power levels equal to 10 DECIBELS.

berkelium (Bk)   Element 97; Atomic wt. 245; Valence 3 or 4. Radioactive with Bk-243 having half-life of 4.6 hours; decays by K-electron capture, forms curium-243.

beryllium (Be)   Element 4; Atomic wt. 9.013; Melting pt. 1350°C; Boiling pt. 1500°C (5 mm); Spec. grav. (20°C) 1.8; Outer electronic configuration $1s^2 2s^2$; Ionic radius (Be^{++}) 0.34Å; Atomic radius (Be^0) 1.11Å; Atomic diameter 2.221Å; Atomic vol. 4.96 cm^3/mole; Crystal structure hexagonal close-packed; Ionization energy (Be$^0\rightarrow$Be^{++}) 27.4 ev; Heat of fusion 250–275 cal/g; Heat of vaporization 53,490 cal/mole; Spec. heat (20–100°C) 0.43–0.52 cal/(g)(°C); Thermal conductivity 0.355 cal/(cm^2)(cm)(sec)(°C); Electrical resistivity (0°C) 4 × 10^{-6} ohm/cm, (100°C) 6 × 10^{-6} ohm/cm; Electrolytic solution potential (Be/Be^{++}) E^0 = 1.69 volts; Electrochemical equivalent 0.04674 mg/coulomb. Valuable in electronics because of its high thermal conductivity and high-frequency electrical insulating properties. Used in the manufacture of high-temperature refractory material, in high-quality electrical porcelains for aircraft sparkplugs, and in ultra-high frequency radar insulators.

beta (*written B, β*)   The second letter of the Greek alphabet, used symbolically to represent any of various constants, coefficients, etc., of which, in electronics, the principal ones are: **1** (*β*) An angle. **2** (*β*) The short-circuit current gain of a transistor amplifier operated in the common-emitter mode. **3** (*β*) Phase constant. **4** (*β*) Beta particle (which see).

beta cutoff frequency   In a transistor, the frequency at which the current amplification produced in the common emitter mode (*β*) drops 3 decibels below the low frequency value, that is, a frequency, f, where

$$\beta(f) = (1/\sqrt{2})\beta_0.$$

beta decay   A radioactive transformation of an atomic nucleus in which the atomic number increases or decreases by 1 while the mass number continues unchanged. This may happen through electron emission, positron emission, or electron capture.

beta particle   An electron or positron emitted by an atomic nucleus during beta decay.

betatron   A circular accelerator in which electrons are confined in their orbits by a steady magnetic field and accelerated by another magnetic field that varies with time. The action is similar to that of a transformer in which the external source of ac power is connected to the primary and the electron beam forms the secondary.

Beverage antenna   A WAVE ANTENNA.

bezel *n.*   A grooved rim that holds a lens, window, graticule, etc., in place.

BFO, bfo *abbr.*   BEAT-FREQUENCY OSCILLATOR.

bi- *prefix*   Two: *bidirectional; bifilar.*

bias *n.*   **1** A steady-state force, voltage, magnetic field, etc., applied to a system or device to establish a reference level or determine the range of operation. **2** *v.* To apply a bias to.

biconical antenna   An omnidirectional antenna consisting of two conical conductors joined vertex-to-vertex along a common axis, with the driving power applied at their junction.

bidirectional *adj.*   Operating in two usually opposite directions.

bifilar *adj.*   Formed of, having, or supported by two wires or threads.

bifilar suspension   A type of galvanometer movement in which a D'Arsonval moving coil is supported at each end by two taut wires. The elimination of the pivot, with its attendant friction, results in superior sensitivity and precision. This form of movement is highly resistant to overloads.

bifilar winding   **1** A method of reducing the inductance of wirewound resistors. The wire is wound onto the form doubled so that the current in any two adjacent turns flows in opposite directions. **2** The winding of the primary and secondary coils of a transformer onto the core at the same time, ensuring very close coupling between them.

bilateral *adj.*   Having a voltage-current characteristic curve that is symmetrical with respect to the origin, that is, being such that if a positive voltage produces a positive current of a certain magnitude, an equal negative voltage produces a negative current of the same magnitude.

bilateral amplifier An amplifier that inherently (that is to say, without external connections made for this purpose) transfers energy from its output circuit to its input circuit.

bilateral antenna An antenna whose response is maximum in two directions that are 180° apart.

bilateral trigger diode A transistor-like device having a symmetrical voltage current characteristic, used principally for triggering thyristors.

billboard antenna An antenna array consisting of several bays of stacked dipoles spaced ¼ to ¾ wavelength apart, with a large reflector placed behind the entire assembly. The required spacing of the dipoles tends to make the array inconveniently large below the VHF range.

bimetallic strip A strip formed of two metals of different rates of expansion, so arranged that the strip deflects when subjected to a change in temperature. Used to control or measure temperature.

binary *adj.* **1** Of or based on 2. A **binary number** is written as a sum of integral powers of 2, each with a coefficient of 0 or 1, that is, in the form

$$\sum_{n=-k}^{+k} m_i 2^n,$$

where m_i is 0 or 1, a **binary digit.** The usual convention of placing the high-order digit at the left is followed; thus binary 1111.01 equals 15.25 in the decimal system. **2** *n.* A bistable circuit.

binder *n.* A thermoplastic material which provides mechanical strength by causing various particles to adhere together to form a solid mass.

binding energy The energy required to remove some constituent from a system, especially an atomic or molecular system.

binding post A terminal consisting of a threaded bolt to which wires may be connected by means of nuts.

binistor *n.* Any of various bistable semiconductors used in switching circuits. See SILICON CONTROLLED RECTIFIER, SILICON CONTROLLED SWITCH, UNIJUNCTION TRANSISTOR.

binomial *n.* An algebraic expression consisting of two terms joined by a plus or minus sign.

binomial array A type of broadside array that can be made unidirectional, in which case the radiation pattern has a single lobe, or bidirectional, in which case there are two lobes. The antennas, spaced at half-wavelength intervals, are fed in the same phase, with their relative current amplitudes following the coefficients of a binomial expansion. See BINOMIAL THEOREM.

binomial theorem In mathematics, a theorem giving the general expansion of a binomial raised to any positive integral power. The theorem states that

$$(a + b)^n = a^n + na^{n-1} b + [n(n - 1)/2!]a^{n-2} b^2 + \cdots$$
$$\frac{n(n - 1)(n - 2)\cdots(n - k + 1)}{k!}a^{n-k} b^k \cdots$$

bionics *n.* The study of animals as "living systems" for the purpose of modeling electronic and other technological systems after them.

bipolar *adj.* Operating by means of current carriers of both polarities. The conventional transistor is a bipolar device.

biradial *adj.* Having an elliptical cross section: used with reference to phonograph styli.

birefringence *n.* DOUBLE REFRACTION—**birefringent** *adj.*

bismuth (Bi) Element 83; Atomic wt. 209; Melting pt. 271.3°C; Boiling pt. 1450°C; Spec. grav. 9.747; Valence 3 or 5; Density (20°C solid) 9.80 g/cm^3, (300°C liquid) 10.03 g/cm^3, (400°C liquid) 9.91 g/cm^3, (600°C liquid) 9.66 g/cm^3; Hardness (Brinell) 4–8; Crystal structure rhombohedral, $a_0 = 4.7457$Å; Heat of fusion 2.51 kcal/mole; Heat of vaporization 42.6 kcal/mole; Mean spec. heat (0–270°C) 0.0294 cal/g. (300°C–1000°C) 0.0373 cal/g; Thermal conductivity (100°C solid) 0.018 cal/(sec) (cm^2)(°C), (300°C liquid) 0.041 cal/(sec)(cm^2)(°C), (400°C liquid) 0.037 cal/(sec)(cm^2)(°C); Coefficient of linear expansion 13.45 $\times$ 10^{-6}/°C; Modulus of elasticity 4.6 $\times$ 10^6 lb/cm^2; Thermal-neutron absorption cross-section 0.032 $\pm$ 0.003 barns. Electrical resistivity (0°C solid) 106.5 $\times$ 10^{-6} ohm/cm, (100°C solid) 160.2 $\times$ 10^{-6} ohm/cm, (269°C solid) 267.0 $\times$ 10^{-6} ohm/cm, (300°C liquid) 128.9 $\times$ 10^{-6} ohm/cm, (400°C liquid) 134.2 $\times$ 10^{-6} ohm/cm, (600°C liquid); Magnetic susceptibility -1.35×10^{-6} cgs. Used in low-melting alloys and special solders.

bistable *adj.* Having two stable states.

bistable multivibrator A multivibrator that has two stable conditions and which changes from one to the other every time it receives an external trigger pulse of one exclusive polarity.

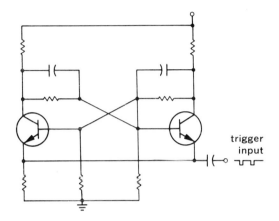

bistatic radar A radar system in which the transmitter and receiver are separately located. In systems of this kind it is possible to have several transmitters and/or receivers.

black body An IDEAL object that absorbs completely all radiation that is incident upon it. At any specified temperature it radiates in each part of the electromagnetic spectrum the maximum energy per unit time available from any object that radiates by virtue of its temperature alone. In practice, the closest approximation of a black body is an almost completely closed cavity in an opaque object.

black box 1 In electrical circuit analysis, a device or network that is considered to be accessible only at its terminals. The device is analyzed in terms of its input and output currents and voltages. This method of analysis is very useful for transistors, electron tubes, and the like, but may be applied to whole circuits or systems. 2 *informal* A unit, usually an electronic module, that can be mounted as a single package.

blacker-than-black In a COMPOSITE VIDEO SIGNAL, a level of voltage past that at which the electron gun of the cathode-ray tube is completely cut off. This range is reserved for the synchronizing and blanking pulses.

blank *v.* To darken the screen of (a cathode-ray tube) by cutting off the electron beam. This is normally done during retrace.

blanking pulse A pulse used to blank a cathode-ray tube.

bleeder current The current drawn by a bleeder.

bleeder or **bleeder resistor** A resistor connected across the output of a power supply. It serves to keep the current drain high enough to make the voltage regulation of the supply adequate, and to allow the filter capacitors to discharge when the equipment is turned off.

blister *n. informal* A RADOME.

block *v.* 1 To prevent the flow of (a current); assume a nonconducting state. 2 To assume a condition wherein all messages cannot be properly routed, as a communications network. See CONGESTION.

block diagram A representation of a system as a number of boxes each labeled as performing some function necessary to the operation of the system, the boxes being connected in the appropriate order by lines.

blocked impedance The electrical impedance of an electromechanical or electroacoustic transducer when it is constrained so that no motion is possible.

blocking *n.* CONGESTION.

blocking capacitor A capacitor connected into a circuit to provide a low impedance path for an alternating-current signal while blocking the flow of direct current.

blocking oscillator A type of feedback oscillator which behaves regeneratively for a short period of time during which the active element (tube, transistor, etc.) is driven into cutoff, remaining there until the associated time constants allow the cutoff bias to decay sufficiently for the action to repeat itself. The output is thus a series of pulses. Compare SQUEGGING OSCILLATOR.

blooming *n.* A condition in which the scanning spot of a cathode-ray tube becomes excessively large, causing considerable distortion of the image or display that is projected.

blowout coil An electromagnet located near contacts that are expected to arc when a circuit is broken. Its function is to deform the arc by means of its magnetic field, thus hastening extinction.

B-minus The negative terminal of a B-supply.

B-negative B-MINUS.

Bode diagram   For a feedback control system or amplifier, a plot of gain against frequency and a plot of phase shift against frequency, used to determine the stability of the system. In a system of this type oscillation occurs at any frequency where gain is greater than or equal to unity and the phase shift between input and output is 180°.

bogey *n.*   An average, published, or nominal value for some characteristic of a device.

bogey electron device   A device whose characteristics match the published nominal characteristics for its type.

Bohr magneton   Either of two units of MAGNETIC MOMENT used in atomic physics. The first, the **electronic Bohr magneton** μ_0 is given by

$$\mu_0 = eh/4\pi m_e c = 9.27 \times 10^{-21}$$

erg per gauss, where h is Planck's constant, e the electronic charge, m_e the rest mass of the electron, and c the speed of light. The second, the **nuclear Bohr magneton** μ_1 is given by

$$\mu_1 = eh/4\pi Mc = \mu_0/1836,$$

where M is the rest mass of the proton and the other symbols are defined as before.

bolometer *n.*   **1** An instrument that measures radiation by means of a metallic detector that changes its resistivity when radiation strikes it and raises its temperature. Instruments of this kind can be made very sensitive. **2** In microwave work, a BARRETTER, THERMISTOR or related device.

Boltzmann's constant   See IDEAL GAS LAW.

bombard *v.*   To expose (substances) to the effect of radiation or to the impact of high-energy particles.

bond *n.*   **1** An electrical attraction uniting two atoms, as in a molecule, crystal, etc. Bonds are attributed to various distributions of the electrons in the VALENCE BANDS of the atoms in question. **2** A low-impedance electrical connection, in particular one made between some shielding element and ground.

Boolean algebra   A mathematical system that can be interpreted as an algebra of logical properties of statements as well as an algebra of sets. It contains operators that are equivalent to AND, OR, NOR, IF, etc. Boolean algebra is used in the designing of digital computers. See LOGIC CIRCUIT, AND GATE, OR GATE, etc.

boom *n.*   **1** A polelike support from which a microphone can be suspended. **2** A spurious bass response produced by a underdamped loudspeaker.

boosted B+ voltage   In a television receiver, the voltage that results from allowing the voltage pulses drawn from the horizontal deflection coil circuit through the damper tube to charge one or more capacitors that are in series with the B+ from the power supply. This voltage, often hundreds of volts above B+, is used as the plate supply for the horizontal output tube.

booster *n.*   **1** In radio broadcasting, a REPEATER. **2** In audio engineering, a line amplifier.

bootstrap amplifier   An amplifier that raises the potential of its input source by an amount equal to the output voltage, or, when as in many applications the gain of the amplifier is unity, raises the potential of the input by an amount equal to the input voltage.

boron (B)   Element 5; Atomic wt. 10.82; Melting pt. 2300°C; Boiling pt. sublimes 2550°C; Spec. grav. (crystals) 3.33^{20}, (amorphous variety) 2.34; Hardness 9.5; Valence 3; Spec. heat 0.309 cal/(g)(°C); Density (25–27°C crystalline) 2.31 g/cm^3, (25–27°C amorphous) 2.34 g/cm^3; Moh hardness (25°C −27°C crystalline) 9.5; Electrical resistivity (25°C) 1.7×10^6 ohm/cm; Coefficient of thermal expansion (20 −750°C) 8.3×10^{-6} cm/°C. Used as an acceptor impurity in semiconductors.

Bosanquet's law   In a magnetic circuit, a law analogous to Ohm's law in an electric circuit, namely that

$$\phi = \frac{\mathcal{F}}{\mathcal{R}},$$

where ϕ is MAGNETIC FLUX, $\mathcal{F}$ is MAGNETOMOTIVE FORCE, and $\mathcal{R}$ IS RELUCTANCE.

bottom *v.*   To reach a point on an operating or characteristic curve where a negative change in the independent variable, as for example the input, no longer produces a constant change in the dependent variable, as for example output.

bow-tie antenna   An antenna consisting of two triangular conductors in the same plane with a reflector behind them. One side of the transmission line is connected to the vertex of each of the triangles. This antenna is used mostly in the UHF band.

boxcar *n.* One of a series of pulses having long duration in comparison to the spaces between them.

boxcar circuit A circuit that during a gated interval allows an input signal to charge a storage capacitor, delivering, through an isolating element such as a cathode follower, an output proportional to the accumulated charge.

B-plus The positive terminal of a B-supply.

B-positive B-PLUS.

branch *n.* In an electrical network, an element or a series connection of two or more elements included between two JUNCTIONS.

brass *n.* An alloy whose principal constituents are copper and zinc, useful for its corrosion resistance.

breadboard 1 *n.* An experimental version of a circuit, generally laid out on a flat board and assembled with temporary connections in order to facilitate substitution of components. **2** *v.* To make a breadboard of.

breakdown *n.* A condition in which an insulator, reverse-biased pn junction, or anything which is normally nonconducting begins to conduct heavily as a result of the application of a strong electric field.

breakdown diode A ZENER DIODE.

break frequency A CORNER FREQUENCY.

break-in operation The operation of a continuous wave transmitter-receiver installation in such a way that the receiver is energized during any interval in which the transmitter is not radiating.

breakover *n.* In a silicon controlled rectifier or related device, a transition into forward conduction caused by the application of an excessively high anode voltage. In some cases this is destructive to the device.

bridge *n.* An electrical network consisting of four impedances. (See Figure). The network has the property that if $Z_1/Z_2 = Z_3/Z_4$ the output voltage is zero; the network is then said to be balanced. If two of the impedances are known, one variable and one unknown, it is possible to measure the unknown by adjusting the variable element until balance is achieved. The network can be arranged so that if the input voltage is AC, the network will be balanced at only one frequency. If, with an AC input, the four impedances are replaced

by rectifiers, the output will be full wave rectified DC. See WHEATSTONE BRIDGE, IMPEDANCE BRIDGE.

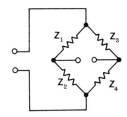

bridged-T (or H, or pi) network A network of the indicated configuration with an additional impedance connected between its input and output.

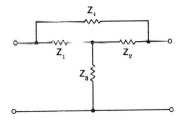

British thermal unit The energy required to raise the temperature of one pound of water one degree Fahrenheit.

broadband 1 *adj.* Having an essentially uniform response over a wide range of frequencies. **2** *v.* To design or adjust (an amplifier) for wide bandwidth.

broadcast *n.* A radio or television transmission intended for general reception.

broadcast band The range of frequencies, between 550 and 1600 kilohertz, assigned to broadcasting stations in the U.S.

broadside array An antenna array whose direction of maximum response lie along a line normal to the plane of the antenna elements.

bromine (Br) Element 35; Atomic wt. 79.916; Melting pt. $-7.2°C$; Boiling pt. $58.78°C$; Density (gas) 7.59 g/l, (20°C liquid) 3.12 g/ml; Valence 1, 3, 5, or 7; Heat of fusion $(-7.2°C)$ 16.14 cal/g; Heat of vaporization $(58.78°C)$ 44.8 cal/g; Critical temp. 311°C; Expansion coefficient (20–30°C) 0.0011 per °C; Spec. heat $(-20.7°C$ crystals) 0.0898 cal/g, (0°C liquid) 0.107

cal/g, (83°C–228°C vapor) 0.0553 cal/g; Electrical conductivity (0–20°C) 10^{-12} mho/cm. Bromine and its compounds are used in storage batteries.

Brown & Sharpe gauge See AMERICAN WIRE GAUGE.

Brownian movement The random movements made by extremely small particles that are suspended in a fluid, now known to be caused by fluctuations of pressure over the surface of the particle due to the random thermal agitation of the molecules of the fluid.

brush *n*. A conductor, generally of graphite, that forms the connection between another rotating conductor and an external circuit.

brush discharge A type of corona discharge occurring in a gas when it is subjected to an electric field strong enough to ionize it in part, but not strong enough to break it down completely. The discharge takes the form of long, hair-like, luminous streamers.

brute force *informal* The use of seemingly inefficient design in order to achieve a desired result. Sometimes this is done in order to avoid involved design procedures, critical adjustments, or the like, but often it is the only possible approach. For example, the miniaturization of low-frequency loudspeakers requires 'brute force' in the form of greatly increased amplifier power.—**brute-force** *adj*.

B-scan A cathode-ray tube display in which the vertical deflection represents distance and the horizontal deflection represents azimuth, used in radar, sonar, and related systems.

B.S.E. Engr. *abbr*. Bachelor of Science in Electrical Engineering.

B & S gauge Brown & Sharpe gauge. See AMERICAN WIRE GAUGE.

B.S.El. Engr. *abbr*. Bachelor of Science in Electronic Engineering.

B supply The source that supplies the operating power for active components such as vacuum tubes, transistors, etc.

BT-cut crystal A slab of quartz cut at an angle of about $-49°$ with the optical axis (Z-axis) of the mother crystal. It has a temperature coefficient of virtually zero.

buck *v*. Act in opposition to; tend to cancel.

buffer *n*. An element whose main function is to prevent undesirable interaction between other elements of a system.

bug 1 *n*. A semiautomatic telegraph transmitting key. The operator produces a series of dots by holding a lever in one position or a continuous dash by holding it in another position. **2** *n. informal* A design or construction flaw in a piece of equipment. **3** *n. informal* A concealed listening device. **4** *v. informal* To equip with a concealed listening device.

bulk eraser A device that erases an entire reel of magnetic tape at one time without the need for passing the tape over an erase head. Generally this is done by subjecting the tape to a strong alternating magnetic field which is gradually reduced to zero.

buncher *n*. The INPUT GAP of a velocity-modulated electron tube.

bunching *n*. VELOCITY MODULATION.

Bunsen cell A primary cell whose negative electrode is a piece of zinc immersed in a sulfuric acid electrolyte and whose positive electrode is of carbon immersed in a depolarizer of concentrated nitric acid, the two liquids being separated by a porous cup. The zinc plate is often amalgamated.

burst signal In television, a COLOR BURST.

bus *n*. **1** A BUS BAR. **2** In digital or similar circuitry, a conductor or system of conductors, that connects any of several sources to any of several destinations.

bus bar A bar or strip of copper or aluminum forming a connection between circuits.

bushing *n*. **1** A metal lining designed to protect machine parts from abrasion. **2** A tube used to reduce the diameter of a hole, pipe, etc. **3** A lining inserted in a socket to insulate an electric current.

Butler oscillator A crystal-controlled oscillator employing two active stages. The crystal is used in its series-resonant mode and acts as the positive feedback path.

butterfly resonator A tuning device similar to an air dielectric variable capacitor except that the stator has a hole cut similar in shape to the wings of a butterfly and the rotor also has this shape. Movement of the rotor varies both the inductance and capacitance (and hence the resonant frequency) of the device. The device is used mainly in the VHF and UHF bands.

button *n.*   **1** In a carbon microphone, the container that holds the carbon granules. **2** A DOT.

button stem   In an electron tube, the glass base through which the pins pass and which supports the structure on which the elements of the tube are mounted.

BWO *abbr.*   Backward-wave oscillator.

bypass 1 *n.*   An element that allows some component of a current to flow around another element while creating a negligible voltage drop. In general the impedance of a by-pass should be equal to 1/10 or less of the impedance of the element it shunts. **2** *v.* To act as or provide a bypass for. **3** *v.* To pass around (an element) by means of a bypass.

bypass capacitor   A capacitor providing a low-imped-

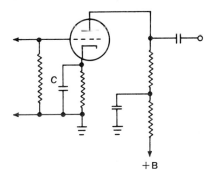

Triode amplifier with the capacitor C bypassing the cathode resistor.

ance path for alternating current around an element in a circuit.

C

C *abbr.*   Coulomb(s).

ca *abbr.*   Cathode.

cable 1 *n.*   One or more wires or transmission lines mechanically assembled into a compact form, often with a protective sheath. **2** *v.* To assemble (several wires, transmission lines, etc.) into a cable.

cadmium (Cd)   Element 48; Atomic wt. 112.41; Melting pt. 320.9°C; Boiling pt. 767°C; Spec. grav. (20°C) 8.65; Hardness 2; Valence 2; Spec. heat 0.055 cal/(g)(°C); Electrical conductivity (0 – 20°C) 0.146×10^6 mho/cm. Used as an electrodeposit coating of iron and steel to protect them against corrosion. Minor uses include the alkaline cadmium-nickel electric storage cell, solder, "bearing metals," and other low melting point alloys. Cadmium sulfate ($3CdSO_4 \cdot 8H_2O$) is used in making Weston standard electromotive force cells.

calcium (Ca)   Element 20; Atomic wt. 40.08; Melting pt. 810°C; Boiling pt. 1170°C; Spec. grav. (20°C) 1.55; Density 1.54; Hardness 1.5; Valence 2+; Electron configuration 2 8 8 2; Ionic radius 0.99Å; Crystal form face-centered cubic; Latent heat of vaporization (1170°C) 399 kilojoules/g-atom; Spec. heat 0.149 cal/(g)(°C); Electrical conductivity (0 – 20°C) 0.218×10^6 mho/cm. Used as a GETTER in electron tubes.

calculus *n.*   **1** The branch of mathematics that analyzes the behavior of FUNCTIONS by means of DERIVATIVES and INTEGRALS. **2** A set of rules and procedures for carrying on various logical or mathematical operations.

calibrate *v.*   **1** To graduate, correct, or adjust the scale of a measuring instrument into appropriate units. **2** To determine the reading of such an instrument. —**calibration** *n.*

californium (Cf)   Element 98; Atomic wt. 246. Radioactive element with isotopes having half-life of a few minutes to 1000 years; emits alpha particles. Used as tracer material in nuclear research.

call letters   A series of letters, or letters and numbers, assigned by some governmental agency to identify a radio or television transmitting station.

calorimeter *n.*   An instrument for measuring changes in the amount of heat contained within a system.

canal rays   In a gas discharge tube, streams of positive ions that are attracted toward the cathode and allowed to pass on through a series of perforations.

candela *n.*   The measure of luminous intensity, defined as equal to 1/60 of the luminous intensity of 1 square centimeter of a BLACK-BODY surface at the temperature of solidification of platinum (1773.5°C). The international standard since 1948.

candle *n.* CANDELA.

candlepower *n.* Luminous intensity as expressed in CANDELAS.

cannibalization *n.* The taking of parts from one machine or system for use in another.

capacitance *n.* **1** The ability of a circuit or circuit element to store energy in the form of electric charge on conductors between which there exists a potential difference. Capacitance is given by $C = Q/V$, where C is capacitance in farads, Q is the stored charge in coulombs, and V is the potential difference in volts. **2** An ideal capacitor.

capacitance bridge An electrical BRIDGE used to determine the ratio of two capacitances.

capacitive *adj.* Of, like, or having to do with capacitance or a capacitor.

capacitive reactance See REACTANCE.

capacitor *n.* A device designed to offer a desired amount of capacitance to a circuit. In general, a capacitor consists of two parallel conductive plates separated by a dielectric. To a good approximation the capacitance (C) of a capacitor is given by:

$$C = 8.84 \times 10^{-8} \, ka/d \text{ microfarads,}$$

where *a* is the area of each plate, *k* the DIELECTRIC CONSTANT of the dielectric, and *d* the distance between the plates.

capacitor microphone A microphone whose active element is a capacitor whose value varies as the sound pressure changes the spacing of its plates. In this way the dc polarizing voltage applied to the capacitor is modulated with an audio signal.

capacity *n.* **1** The amount of electricity that can be drawn from a cell or battery, usually expressed in ampere-hours. **2** Capacitance.

capstan *n.* In a tape recorder, a rotating spindle against which the tape is held in such a way that it is pulled past the heads at a constant speed during recording and playback.

capture effect The ability of an FM receiver to suppress the weaker of two signals that arrive on the same or close to the same frequency. It can be shown that the weaker signal in effect amplitude modulates the stronger. As FM receivers will reject amplitude modula-

tion (provided that it is not too great a percentage of the total signal), the weaker carrier is suppressed.

capture ratio The minimum ratio of signal strengths for which the capture effect will completely suppress the weaker signal.

carbon (C) Element 6; Atomic wt. 12.010; Melting pt. 3550°C, sublimes above 3500°C; Boiling pt. 4200°C; Density 2.26 g/ml; Hardness (diamond) 10, (graphite) 0.5−1.0. Spec. grav. (amorphous) 1.88, (graphite) 2.25, (diamond) 3.51; Valence 2, 3, or 4; Spec. heat 0.165 cal/(g)(°C); Electrical conductivity (0−20°C) 0.7×10^3 mho/cm. In its graphitic form it is used for high-temperature crucibles, dry-cell and arc-light electrodes. It is also used in manufacturing resistors.

carbon microphone A microphone whose active element is a container filled with carbon granules. The variations in pressure caused by the sound waves are coupled to this container which varies in resistance accordingly.

carbon steel An alloy of iron, 0.9% carbon and 1% manganese used for permanent magnets, having a coercive force of 50 oersteds and a remanence of 10,000 gauss.

carcinotron *n.* A type of backward-wave oscillator tube, tunable from UHF to over 100 ghz. See BACKWARD-WAVE TUBE.

cardioid *n.* The curve formed by the polar plot of the function

$$r = a(1 - \cos \phi).$$

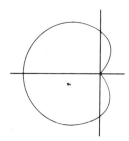

cardioid microphone A microphone whose directivity pattern takes the form of a CARDIOID.

Carey-Foster bridge A form of WHEATSTONE BRIDGE adapted for the comparison of very nearly equal resistances. Provision is made for the elimination of stray resistances from imperfect contacts.

Carpenter Temperature Compensator 30   BALCO.

carrier *n.*   **1** A flow of energy that can be varied (modulated) in such a way as to carry INFORMATION. See MODULATION. **2** A mobile concentration of electric charge, as for instance an electron, or a hole.

carrier wave   A wave, usually a radio wave, that functions as a carrier (def. 1).

cascade *n.*   A connection of TWO-PORT NETWORKS in which the output of one network becomes the input of the next.

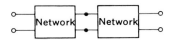

Two four-terminal networks connected in cascade.

cascode amplifier   A cascade amplifier consisting of a grounded cathode input stage driving a grounded grid output stage. Frequently used at the input of UHF and VHF receivers because of its low noise figure.

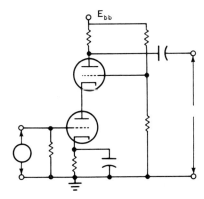

Cassegrain feed   A method of feeding a reflector antenna in which a waveguide located in the center of the main reflector feeds energy to a small reflector which reflects it in turn to the main reflector.

catcher or **catcher resonator**   In a klystron, an OUTPUT RESONATOR.

catelectrotonus *n.*   The increased sensitivity produced in a nerve or muscle in the region of contact with the cathode when an electric current is passed through it.

catenoidal horn   A HYPERBOLIC EXPONENTIAL HORN.

cathode *n.*   The electrode through which current leaves a device, that is, the negative terminal of a device that uses electricity, or the positive terminal of a device that produces electricity. See ANODE.

cathode dark space   In a glow discharge tube, the dark region between the cathode and the luminous region.

cathode follower   An electron-tube amplifier in which the output load is connected in the cathode circuit and the input signal is applied between the control grid and ground. The circuit is characterized by heavy NEGATIVE FEEDBACK, and thus, while the limit of its voltage gain is unity, it has high input impedance, low output impedance, and excellent linearity. Despite the low voltage gain the low output impedance does permit power gain. Its principal use is as an isolation (buffer) stage between otherwise incompatible cascaded circuits.

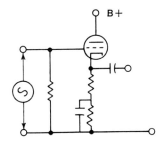

cathode ray   A stream of electrons, as those emitted by a thermionic cathode.

cathode-ray tube   An electron tube in which a stream of electrons is emitted, formed into a beam, accelerated,

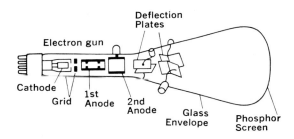

and deflected so as to impinge on various parts of a phosphor-coated screen, forming a luminous display.

cathodoluminescence *n.* Luminescence that results from bombardment with cathode rays.—**cathodoluminescent** *adj.*

cathodophosphorescence *n.* Phosphorescence that results from bombardment with cathode rays.—**cathodophosphorescent** *adj.*

cation *n.* The positive ion of an electrolyte.

cavitation *n.* The production of gas-filled cavities in a liquid when the pressure is reduced below a certain critical value with no change in the temperature. Ordinarily this is a destructive effect as the high pressures produced when these cavities collapse often damage mechanical components of hydraulic systems. In ULTRASONIC CLEANING, however, the effect is often turned to advantage.

cavity or **cavity resonator** A space completely bounded by a metallic conductor. Such a space acts as a resonant chamber for electromagnetic radiation, the resonant frequency or frequencies depending on the geometry of the space.

C band A band of microwave frequencies extending from 3.9 ghz (3.9×10^9 hertz) to 6.2 ghz.

cc, c.c. *abbr.* Cubic centimeter(s).

CCIF *abbr.* International Telephone Consultative Committee.

CCIR *abbr.* International Radio Consultative Committee.

CCIT *abbr.* International Telegraph Consultative Committee.

CD *abbr.* Capacitor-diode.

cell *n.* Any or various single units that produce electricity from either chemical energy, radiant energy, or heat. See ELECTROLYTIC CELL, PHOTOCELL, etc.

cell-type tube Any of various gas-filled, radio frequency switching tubes that are used in conjunction with resonant cavities. See ANTITRANSMIT-RECEIVE TUBE, PRE-TRANSMIT-RECEIVE TUBE, TRANSMIT-RECEIVE TUBE.

Celsius scale A temperature scale in which the freezing point of water at normal atmospheric pressure is 0° and the boiling point is 100°.

cent *n.* A measure of frequency, defined as equal to 100th of a semitone.

centi- *combining form* One hundredth (10^{-2}) of a specific quantity or dimension.

Centigrade scale The CELSIUS SCALE.

ceramic *adj.* Pertaining to or made of clay or other silicates.

cerium (Ce) Element 58; Atomic wt. 140.13; Melting pt. 640°C; Boiling pt. 1400°C; Spec. grav. (20°C) 6.90; Density 6.8; Valence 3 or 4; Spec. heat 0.042 cal/(g) (°C); Electrical conductivity (0–20°C) 0.013 $\times$ 10^6 mho/cm.

cesium (Cs) Element 55; Atomic wt. 132.91; Melting pt. 28.5°C; Boiling pt. 670°C; Spec. grav. (20°C) 1.873; Valence 1; Heat of fusion (28.5°C) 3.8 cal/g; Heat of vaporization (705°C) 146 cal/g; Viscosity (100°C) 4.75 millipoises; Vapor pressure (278°C) 1 mm, (635°C) 400 mm; Thermal conductivity (28.5°C) 0.044 cal/(sec)(cm²)(°C); Heat capacity (28.5°C) 0.06 cal/(g)(°C); Electrical resistivity (30°C) 36.6 $\times$ 10^{-6} ohm/cm. Used in photoelectric cells, in military infrared signaling lamps, and as a GETTER in vacuum tubes.

cfd *abbr.* Cubic feet per day.

cfh *abbr.* Cubic feet per hour.

cfm, c.f.m. *abbr.* Cubic feet per minute.

cfs, c.f.s. *abbr.* Cubic feet per second.

C.G.S. *abbr.* Centimeter, gram, second.

C.G.S. (or **cgs**) **system** The system of measure in which the basic unit of length is the centimeter, the basic unit of mass is the gram, and the basic unit of time is the second.

chaff *n.* A mass of metallic foil strips dropped from military aircraft in order to saturate enemy radar with a multiplicity of spurious reflections.

channel *n.* **1** Something that forms a path for a signal, in particular, a band of radio wave frequencies reserved for some form of communication. **2** The conduction path between the SOURCE and DRAIN of a FIELD EFFECT TRANSISTOR or a related device. **3** In a computer, the part of a data storage that is accessible to a given reading station. **4** In a digital computer, a unit controlling one or more input/output devices.

characteristic *n.* **1** A significant property of a device or system, measured in terms of related parameters, as current and voltage, force and flux, etc. Characteristics are often plotted as curves or families of curves. **2** The integral part of a LOGARITHM to the base 10; also, the power of 10 by which the significant digits of a floating-point number are multiplied. See FLOATING-POINT ARITHMETIC.

characteristic distortion Distortion that depends not only on the properties of the transmission system but also on the previous history of the system, that is, distortion that results from transients caused by earlier signals or from remnants of earlier signals.

characteristic equation A polynomial equation that represents a homogeneous, linear DIFFERENTIAL EQUATION in operational form, that is, given a differential equation of the form

$$A_n \frac{d^n x}{dt^n} + A_{n-1} \frac{d^{n-1} x}{dt^{n-1}} \cdots + A_0 = 0,$$

where $x(t)$ is a solution, the characteristic equation is

$$A_n s^n + A_{n-1} s^{n-1} \cdots + A_0 = 0,$$

where s represents the operator

$$\frac{d}{dt}$$

See CHARACTERISTIC ROOTS.

characteristic frequency See CYCLOTRON FREQUENCY, COLLISION FREQUENCY, LARMOR FREQUENCY, PLASMA FREQUENCY.

characteristic impedance The impedance (Z_0) that when coupled to the output of a TRANSMISSION LINE absorbs all the power from the line without reflection. For a transmission line having a series inductance L per unit length, a series resistance R per unit length, a shunt capacitance C per unit length, and a shunt conductance G per unit length, the characteristic impedance is given by

$$Z_0 = \sqrt{\frac{R + j\omega L}{G + j\omega C}} = \sqrt{\frac{R + j(2\pi f L)}{G + j(2\pi f C)}}$$

See DISTORTIONLESS LINE.

characteristic roots The ROOTS, in general COMPLEX NUMBERS, of a CHARACTERISTIC EQUATION, equal in number to the degree of the characteristic equation (and to the ORDER of the differential equation it represents). The form of the solution of the differential equation

is uniquely determined by these roots. If the roots are distinct, the solution $x(t)$ is given by

$$x(t) = K_1 e^{s_1 t} + K_2 e^{s_2 t} + \cdots + K_n e^{s_n t},$$

where $s_1, s_2, \ldots s_n$ are the roots and $K_1, K_2, \ldots K_n$ are constants determined by the initial conditions. In the case of multiple roots $x(t)$ is given by

$$x(t) = K_1 e^{s_1 t} + K_2 t e^{s_1 t} + \cdots K_q t^{q-1} e^{s_1 t},$$

where q is the multiplicity of the root and $K_1, K_2, \ldots K_q$ are constants determined from initial conditions. These are the principal cases arising in the analysis of linear systems. See COMPLEX EXPONENTIAL.

characteristic wave impedance A property of a wave medium that is analogous to the CHARACTERISTIC IMPEDANCE of a transmission line. Characteristic wave impedance (η) is given for an electromagnetic wave by

$$\eta - \sqrt{\frac{\mu}{\varepsilon}},$$

where μ is the PERMEABILITY and ε the PERMITTIVITY of the medium.

charge 1 n. A quantity of electricity, expressed in units of COULOMBS. The basis for a quantitative measure of charge is the electron, having a charge of 1.6×10^{-19} coulomb. **2** n. An observed property of certain bodies whereby they exert forces on each other independently of gravitation. **3** v. To take a charge on. **4** v. To apply a charge to. See COULOMB'S LAW.

charge density The charge per unit volume in some region.

chaser n. An array of elements similar to a RING COUNTER except that as each successive element is switched to the 'on' condition the others remain on as well; when all stages are on the next pulse turns them all off.

chassis n. The metal structure that serves as a mount for the components of a piece of electronic equipment.

chatter n. Repeated opening and closing of electrical contacts after they have been brought together. In general, this is an undesirable effect.

cheese antenna An antenna that uses a reflector that has the form of a parabolic cylinder enclosed by plates perpendicular to its elements. The plates are spaced so as to permit no more than one MODE to propagate in the desired direction of polarization. The antenna is fed on the focal line.

chemi- *combining form*   Of, resulting from, or having to do with chemical action: *chemiluminescence*.

chi *n.* (*written X, x*)   The twenty-second letter of the Greek alphabet, used symbolically to represent any of various coefficients, constants, etc., of which in electronics the principal one is (x) SUSCEPTIBILITY.

Child-Langmuir-Schottky law   An empirical law stating that the current (*i*) in a thermionic vacuum diode is given by

$$i = Ge_b^{3/2},$$

where *G* is the PERVEANCE and e_b is the plate voltage. In modified form the law applies also to multi-element tubes.

chirp *n.*   A variation in the frequency of a continuous wave carrier as it is keyed. This is an undesirable effect.

chirp radar   A radar system in which the operating frequency is swept during the time that a pulse is being transmitted. The returning echo is compressed by being passed through a network that introduces a delay proportional to frequency before being displayed. The principal advantages of the system are increased power efficiency and better immunity to noise and jamming.

chlorine (Cl)   Element 17; Atomic wt. 34.457; Melting pt. −101.6°C; Boiling pt. −34.6°C; Density 3.214 g/ml; Spec. grav. (−33.6°C) 1.56; Valence 1, 3, 5, or 7; Critical temp. 144°C; Critical pressure 76 atm; Spec. heat 0.116 cal/(g)(°C).

choke *n.*   An inductor whose primary function is to present a high impedance to an alternating current.

chopper *n.*   A device or system for interrupting a flow of current, light, or other energy at regular intervals. Choppers are often used to make possible the amplification of small dc potentials without the use of direct-coupled amplifiers, which are inherently too unstable for this application.

chroma *n.*   The distinctive quality, excluding brightness, that identifies any particular color. Black, white, and gray, differing only in brightness, lack chroma.

chromaticity diagram   A diagram in which colors are represented as points on a coordinate plane whose

axes are two of the primary colors. Specification of the value of the third primary is unnecessary as the sum of the three equals unity. The specification of a color in this way is called its **chromaticity**.

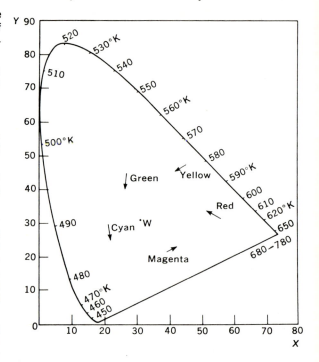

chromatron *n.*   A color kinescope that has a single electron gun and whose color phosphors are laid out in parallel lines on its screen. The electron beam is directed to the correct phosphor by a deflection grid or wire grille near the face of the tube.

Chromel C   An alloy of 24% iron, 16% chromium and 60% nickel having high resistivity, used in rheostats.

Chromel P   An alloy of 90% nickel and 10% chromium, used with ALUMEL in high temperature thermocouples.

chrominance *n.*   The difference between two different but equally bright colors.

chromium (Cr)   Element 24; Atomic wt. 52.01; Melting pt. 1615°C; Boiling pt. 2200°C; Spec. grav. (20°C) 7.1; Hardness 9; Valence 2, 3, or 6; Electrical resistivity (20°C) 13.0 × 10⁻⁶ ohm/cm for annealed and electrodeposited chromium; Heat capacity (25°C) 5.55 cal/(mole)(°C), (liquid) 9.40 cal/(mole)(°C); Latent heat of fusion 4.20 cal/mole; Latent heat of vaporization 76,635 cal/mole; Vapor pressure (2097°C) 10^{-1} atm; Thermal conductivity (20°C) 0.16 cal/(sec)(cm)(°C). Used in electroplating, and in high-resistance alloys for heating units.

chromophore *n.*   Any of various groups of atoms that cause organic compounds to absorb light in characteristic ways.

chronometer *n.*   A timekeeping instrument of high precision for use in navigation and scientific work.

chronopher *n.*   An instrument used to signal the correct time to distant points electrically.

chronoscope *n.*   An instrument for measuring a minute interval of time.

CIE *abbr.*   Commission Internationale de l'Eclairage (International Commission on Illumination).

circ. *abbr.*   Circuit.

circuit *n.*   An electrical network in which there exists one or more paths through which current can flow.

circuit breaker   A switching device that automatically disconnects a source of electric power from its load when an excessively large current is drawn for too long a time.

circular functions   The TRIGONOMETRIC FUNCTIONS. See also HYPERBOLIC FUNCTIONS.

circular mil   A measure of area used in specifying the cross sections of conductors, defined as equal to the area of a circle whose diameter is 1 MIL.

circular polarization   See POLARIZATION.

circulator *n.*   A microwave network arranged so that energy can flow only from each terminal to the next one in sequence. This is accomplished by the use of one or more GYRATORS.

citizens' band   Either of the radio-frequency bands reserved for the citizens' radio service, namely those at 460 Mhz–470 Mhz and 27.23 Mhz–27.28 Mhz.

citizens' radio service   A radio service intended for private and personal communications, radio signaling, remote control of objects, etc., but specifically not intended to include radio communication as a hobby. It is subject to licensing by the Federal Communications Commission.

clamp *n.*   In a vacuum tube, a condition in which due to the flow of grid current the grid is held at cathode potential.

clamper *n.*   A device that produces a desired reference voltage at some point in a network, usually operating in such a way that a waveform appearing at that point has one of its extremes at the reference level.— **clamp** *v.*

Clapp oscillator   A Colpitts oscillator modified so that it is tuned by a series LC circuit, exhibiting as a result improved stability of frequency.

Clark cell   An older form of standard cell that produces 1.433 volts at 15°C. The cathode is of zinc amalgam, the anode of mercury, and the electrolyte of zinc and mercurous sulfates.

class-A signal area   An area in which television signals are available at a level of 2,500 microvolts per meter or more for VHF and 5,000 microvolts per meter or more for UHF.

class (of amplifiers)   Any of a set of designations that describe vacuum-tube and transistor amplifiers in terms of their operating conditions. In a **class-A** amplifier the operating point is chosen so that output current flows during all parts of each cycle of the input signal; in **class-AB** output current flows for more than half but less than all of each cycle; in **class-B** output current flows for approximately half of each cycle; in **class-C** output current flows for less than half of each cycle. In the descriptions of vacuum-tube amplifiers a subscript numeral is often added to the class designation, a 1 indicating that grid current never flows, a 2 indicating that grid current does flow for some part of each cycle. Generally, to avoid distortion, class-AB and B stages are operated push-pull. If a single frequency is to be amplified a class-B or C stage may be operated singled-ended into a tuned load.

class (of broadcast stations)   A set of designations that classify broadcast stations in terms of authorized service and power. A **class I station** serves a PRIMARY and SECONDARY SERVICE AREA, using a CLEAR CHANNEL and an operating power between 10 and 50 kw. A **class II station** serves a primary service area, using a clear channel and an operating power between 0.25 and 50 kw. A **class III** station is a regional station with a primary service area. Its operating power may be between

1 and 5 kw **(class III-A)** or between 0.5 and 1 kw by night and up to 5 kw by day **(class III-B).** A **class IV station** is a local station limited to an operating power between 0.1 and 0.25 kw.

class (of modulators)   Any of a set of designations that describe modulators in terms of the class of amplifier that supplies the signal power used to modulate the carrier.

clear *v.*   In a digital computer, to replace each of the digits in (a register or storage location) by a zero.

clear channel   In the standard broadcast band, a channel such that the station assigned to it is free of objectional interference through all of its PRIMARY SERVICE AREA and most of its SECONDARY SERVICE AREA.

Clichier metal   An alloy of 33% lead, 47% tin, 9% bismuth, and 11% antimony, used in fuses.

click filter   A filter, usually of the resistance-capacitance type, used to suppress the transient that occurs when a switch is closed or opened.

clock *n.*   A pulse generator that is very stable with respect to frequency, pulse duration, and pulse amplitude, used to provide synchronization or a time base for a digital system.

closed *adj.*   Complete; unbroken, as a circuit or loop.

closed-circuit television   A television system in which the signal is not broadcast but is delivered to the associated receivers by cable or some other means of private transmission.

close-talking *n.*   The practice of holding a microphone close to the lips when speaking, often in order to discriminate against ambient noise.

cloverleaf *n.*   An antenna in which each element consists of four loops joined in a way that creates a resemblance to a four-leaf clover. Often, for VHF transmission, these elements are used in a vertical stack separated by half wavelengths.

clutter *n.*   In a radar system, undesirable echos from the ground, the sea, rain, jamming, etc., that tend to make the echos that are of interest less discernible.

cm. *abbr.*   Centimeter(s).

CML   CURRENT-MODE LOGIC.

coax *n.*   Coaxial cable.

coaxial *adj.*   Having a common axis.

coaxial cable or **coaxial line**   A transmission line in which one conductor is completely surrounded by the other, the two having a common longitudinal axis and being separated by some form of dielectric. Such a line produces no external field and cannot be affected by an external field.

coaxial loudspeaker   A loudspeaker consisting of two radiators, generally a woofer and a tweeter, mounted on the same central axis as parts of the same mechanical structure.

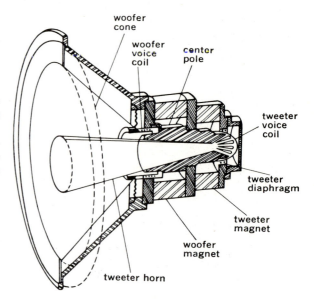

woofer cone — woofer voice coil — center pole — tweeter voice coil — tweeter diaphragm — tweeter magnet — woofer magnet — tweeter horn

cobalt (Co)   Element 27; Atomic wt. 58.94; Melting pt. 1495°C; Boiling pt. 3000°C; Spec. grav. (20°C) 8.9; Valence 2 or 3; Electron configuration $1s^2 2s^2 2p^6 3s^2 3p^6 3d^7 4s^2$; Oxidation states 2+,3+; Spec. heat 0.099 cal/(g)(°C); Electrical conductivity (0−20°C) 0.16 × 10^6 mho/cm. Used in electroplating because of its hardness and resistivity.

codan *n.*   A carrier-operated SQUELCH CIRCUIT.

code *n.*   A system of symbols and the rules by which they may be used to express information.

coefficient *n.*   **1** In general, the product of all but a specified set of the factors of an algebraic term, but most commonly the product of all the constant factors.

2 A number, often a constant, that expresses some property of a physical system in a quantitative way.

coercive force   COERCIVITY.

coercivity *n.*   In a ferromagnetic material, the magnetizing force needed to reduce the residual magnetic flux to zero after the material has been magnetized to saturation.

cogging *n.*   Irregular speed of rotation, especially in an electric motor.

coherence *n.*   The property of a set of waves by which their phases are completely predictable along an arbitrarily specified surface in space, also, the relation between a set of sources by which the phases of their respective radiations are similarly predictable.—**coherent** *adj.*

coho *n.*   A coherent oscillator.

coil *n.*   A number of turns of wire, usually wound onto some form of core but often self-supporting, generally designed to act as an INDUCTOR.

coincidence circuit   An AND-GATE.

cold cathode   In an electron tube or gas discharge tube, a cathode that operates without the need for being heated by an external source of power.

cold weld   A mechanical joint between two pieces of metal, produced by the application of high pressure, but without heat.

colidar *n.*   An optical system similar to radar, operating by means of the light from a laser.

colinear antenna   An antenna array composed of several half-wave dipoles arranged end to end and operating all in phase.

collector *n.*   **1** In a TRANSISTOR, the region into which majority carriers flow from the BASE under the influence of a reverse bias across the two regions. The collector is analogous to the PLATE of a vacuum tube. **2** The external terminal of a transistor that is connected to this region. **3** In certain electron tubes, an electrode to which electrons or ions flow after they have completed their function.—**collect** *v.*—**collection** *n.*

collimator *n.*   An apparatus that adjusts a beam of light so that all its rays are parallel; by extension, a device that performs the same function for any form of radiation.—**collimate** *v.* —**collimation** *n.*

collinear array   An antenna array consisting of a number of half-wave elements arranged on common longitudinal axis in the vertical or horizontal plane.

collision *n.*   An interaction between bodies or particles in which they come close enough together to influence each other, generally with some exchange of energy. In this process the momentum of the system is always conserved; if the energy of the system is conserved as well, the process is called an **elastic collision.**

collision frequency   The average rate at which particles collide. It may appear in the force equation for a particle in the following way:

$$m\,dv/dt = f - m\nu_c v,$$

where f is force, m mass, v velocity, and ν_c the collision frequency.

color burst   In color television, a burst of about 8 cycles of the color carrier transmitted on the BACK PORCH of the horizontal sync pulses, used as a reference in controlling the frequency and phase of the COLOR OSCILLATOR.

color code   A method of indicating the values and ratings of electronic components by various colored stripes, dots, etc., on their outer coverings. See appendix p. 227.

colorimetry *n.*   The measurement and analysis of colors, either by comparison with standards or in terms of physical parameters.

color oscillator   In a color television receiver, the local crystal-controlled oscillator that generates the 3.579545 Mhz subcarrier needed to demodulate the color information. See COLOR BURST.

color temperature   The temperature, in degrees Kelvin, to which a BLACK BODY must be heated in order that its radiation match the color being measured.

Colpitts oscillator   A sinusoidal oscillator using a three-terminal active element, as a tube, transistor, etc., and a feedback loop containing a parallel LC circuit. The capacitance of the LC circuit consists of two capacitors in series forming a voltage divider that serves to match the input and output impedances of the active device.

column loudspeaker   An array of loudspeakers arranged in a vertical line, having the property of spreading its radiation through a wide angle in the horizontal plane while keeping it in a beam with respect to the vertical plane.

comb filter   A filter that has a series of narrow pass bands separated by a series of narrow stop bands. One form of comb filter consists of a delay line in a feedback loop.

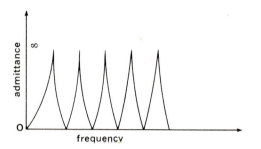

Typical comb filter response

combination tones   Any of the additional audio frequencies produced when two or more frequencies interact in a nonlinear medium or transducer. See INTERMODULATION.

common 1 *adj.*   Shared by two or more circuits. *Common* in this sense is often used to designate the terminal of a three-terminal device that is shared by the input and output circuits. Thus a transistor may be operated in a *common*-base configuration, a *common*-collector configuration, or a *common*-emitter configuration. Vacuum-tube connections may be characterized in a similar way, but *grounded* is normally used instead of *common*. **2** *n.* A point that acts as the reference potential for several circuits; a ground.

common-mode rejection   A figure of merit that expresses the ability of a differential amplifier or measuring device to suppress a signal applied simultaneously to both of its inputs.

common-mode signal   A signal that appears at both inputs of a DIFFERENTIAL AMPLIFIER at the same time.

communication theory   STATISTICAL COMMUNICATION THEORY.

commutation *n.*   The switching of currents back and forth between various paths as required for operation of some system or device, in particular: **1** The switching of current to or from the appropriate armature coils of a motor or generator. **2** The turning off of an active element at the correct time, as in an inverter or power controller.

commutator *n.*   A device that switches the current to or

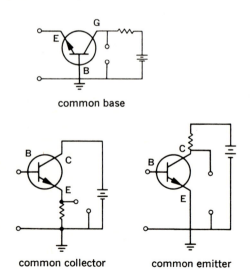

common base

common collector   common emitter

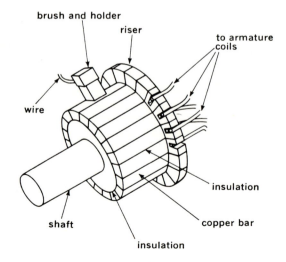

from the moving coils of a motor or generator as required for proper operation.

compander *n.* An electronic device that combines the
functions of a COMPRESSOR and a complementary EX-
PANDER.

comparator *n.* A circuit that compares a changing voltage to a reference voltage and delivers an output
signal when the two are the same.

comparison bridge A BRIDGE used to compare the ratio
of two essentially similar IMPEDANCES.

compensation *n.* The introduction of additional components into a system in order to suppress or minimize
various parasitic effects, in particular: **1** The use of
extra reactive elements in an amplifier, attenuator,
etc., in order to balance out the effects of stray and/or
unavoidable reactances. **2** The use of temperature
sensitive elements in conjunction with semiconductors,
etc., in order to stabilize operating points.

compiler *n.* A PROGRAM that allows a computer to
translate a source program written in something other
than MACHINE LANGUAGE into an object program in machine language.

complement *n.* Either of two numbers which are
derived by rule from a given number expressed to the
base (radix) N and which are often useful in representing the negative of the given number. If the digits of
the given number are $X_1, X_2, \ldots X_n$, the corresponding
digits of the **radix-minus-one** $(N-1)$ **complement** are
$(N-1) - X_1, (N-1) - X_2, \ldots (N-1) - X_n$. If 1 is
added to the radix-minus-one complement, all carries
being performed in the usual manner, the result is the
radix (N) **complement.** If K' is the N complement of K,
then $L + K' = L - K$, provided that in the addition carries out of the high-order position are dropped. A simple example should suffice. Consider (555 radix 10); its
$N - 1$ (9) complement is 444; its N (10) complement
is 445. Now, adding 555 and 445 we get 1000; dropping the high-order carry we get 000 or $555 - 555$.
This method permits great simplification in the hardware of digital computers.

complementary color Either of a pair of spectrum
colors that when combined give a white or nearly white
light.

complementary symmetry The operation of a pnp
transistor and an npn transistor in push-pull with the
same input signal. This arrangement inherently pro-

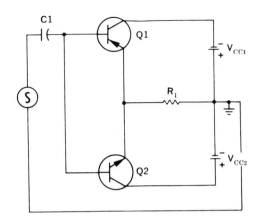

Transistor totem pole amplifier
using emitter followers in complementary symmetry.

duces the necessary phase inversion without requiring
an extra stage.

complementary wavelength Of a color sample, the
wavelength of light that produces white when mixed
with the sample in appropriate proportions. Some colors, as various purples, lack DOMINANT WAVELENGTHS
and are, therefore, specified by their complementary
wavelengths.

complex exponential A COMPLEX NUMBER e^{st} used as a
convenient representation of damped or undamped
sinusoidal oscillations, where

$$s = -\alpha \pm j\omega_d,$$

and e is the base of the natural logarithms. Expansion
of the exponential function gives

$$e^{st} = e^{-\alpha t} \cdot e^{\pm j\omega d t}.$$

The real parts of e^{st} (hereafter called Re $\{e^{st}\}$), is given
by a function of time

$$F(t) = e^{-\alpha t} \cos \omega_d t,$$

where ω_d is the **damped radian frequency** and α is the
attenuation factor. (Note that $e^{-\alpha t} \leq 1$ for $\alpha \geq 0$.) The
response of a second-order system containing resist-
ance (R), inductance (L), and capacitance (C) is a
damped sinusoid given by

$$F(t) = \text{Re}\,\{e^{st}\},$$

where

$$\alpha = -R/2L,$$

and

$$\omega_d = \sqrt{\frac{1}{LC} + \alpha}.$$

It is usual to call

$$\sqrt{\frac{1}{LC}} = \omega_0$$

the natural frequency of the system. See DAMPING FACTOR, CRITICAL DAMPING; OVERDAMPED; UNDERDAMPED; CHARACTERISTIC EQUATION.

complex frequency   See COMPLEX EXPONENTIAL.

complex number   A number of the form $a + bi$, where a, b are real numbers and $i = \sqrt{-1}$. Often in physical and electrical problems the symbol j replaces i. $z = x + iy$ is the definition of a **complex variable.** If we define

$$\rho = \sqrt{a^2 + b^2},$$

and

$$\phi = \tan^{-1} b/a,$$

then

$$a + bi \equiv \rho\,(\cos \phi + i \sin \phi) \equiv \rho e^{\phi},$$

where e is the base of the natural logarithms.

complex plane   A system of rectangular (Cartesian) co-ordinates in which the real part of a COMPLEX NUMBER is plotted on the abscissa and the imaginary part on the ordinate.

complex power   A COMPLEX NUMBER

$$\overline{S} = P + jQ = |\overline{S}|\,e^{j\theta}$$

similar to a PHASOR, used to describe the rate at which energy is both stored and dissipated within a network in the sinusoidal steady state. The real part of the complex power $\text{Re}[\overline{S}] = P$ (often called **average power** and measured in WATTS) gives the rate at which energy is dissipated; the imaginary part $\text{Im}[\overline{S}] = Q$ (often called **reactive power** and measured in VARS) gives the rate at which energy is stored. The magnitude $|\overline{S}|$ is called

apparent power and is measured in VOLT-AMPERES. The angle $\theta = \tan^{-1}(Q/P)$ is called the phase angle of the network. Since $\text{Re}[\overline{S}] = |\overline{S}| \cos \theta$, $\cos \theta$ is called the power factor and θ often, the power factor angle of the network. It follows that in a purely resistive network $\overline{S} = P$ and $\cos \theta = 1$, and, that in a purely reactive network $\overline{S} = Q$ and $\cos \theta = 0$. Thus in the first case all the energy is dissipated, while in the second all the energy is stored. See IMPEDANCE.

complex variable   A variable that is a COMPLEX NUMBER. See FUNCTION.

complex wave   A wave that contains two or more sinusoidal components.

compliance *n.*   In acoustics and mechanics, the reciprocal of STIFFNESS. Compliance is analogous to electrical capacitance.

component *n.*   **1** Any of a set of independently variable elements, as vectors, voltages, forces, etc., whose linear combination determines the overall response of a physical system. **2** A constituent part.

composite circuit   A multiplex transmission circuit used for telegraphic and voice communications, the two being alloted different frequency bands.

composite video signal   The complete video signal as it leaves the transmitter of a broadcasting station, consisting of a luminance signal, blanking pulses, and synchronizing pulses. A color signal contains in addition the sidebands of the color subcarrier and a color burst signal. (See diagram at bottom of page 35.)

compound connection   A method of connecting two or more transistors so that they behave, in effect, like one

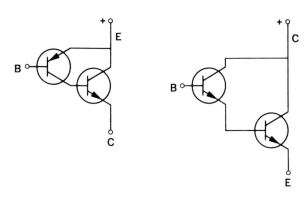

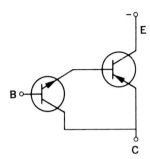

transistor whose current gain (α) is the product of the individual current gains. The limitation of this method is that the leakage currents are amplified as well.

compress *v.* To reduce some parameter of (a signal), as bandwidth, amplitude variation, duration, etc., while preserving its information content.

compressor *n.* A device that compresses a signal, in particular, an amplifier whose gain is automatically ad-justed according to the signal level in such a way that weak signals are amplified more strongly than strong signals, the purpose being to reduce the range of signal strengths. In order to avoid creating distortion, a com-presser acts relatively slowly.—**compression** *n.*

computer *n.* A machine capable of accepting data and processing it by the application of mathematical and/or logical operations. See ANALOG COMPUTER, DIGITAL COMPUTER.

condenser *n.* A CAPACITOR.

condenser microphone A CAPACITOR MICROPHONE.

conductance *n.* In an electric circuit, the reciprocal (G) of RESISTANCE.

conduction band A range of electron energy levels in which the electrons are mobile and hence able to act as current carriers. The concept is most useful in an-alyzing the electrical behavior of materials, including metals, semiconductors, and insulators

conduction current See CURRENT.

conductivity *n.* The degree to which a material will

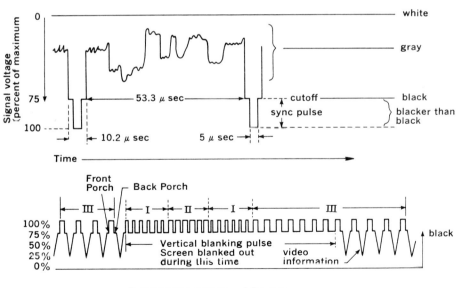

I—Equalizing pulses — 2.5 μ sec
II—Vertical sync pulses — 190.5 μ sec
    Serrations in vert sync pulses 4.4 μ sec
III—Horizontal sync pulses — 5 μ sec

conduct electricity, defined as the conductance between the opposite faces of a cubic centimeter of the material.

conductor *n.* A material that has a low electrical RESISTIVITY, in particular, a wire or bar made of such a material, used to convey current between two points.

conduit *n.* A tubular passage or covering for electrical wires or cables.

conelrad *n.* A system for providing radio service during an air attack in such a way that enemy aircraft will not find it possible to home in on the transmitters.

confidence *n.* The probability that the value of a certain parameter lies between two stated limits. The interval defined by these limits is called the **confidence interval.**

congestion *n.* A condition in which the number of calls arriving at the various inputs of a communications network are too many for the network to handle at once and are subject to delay or loss. The concept applies in an analogous way to any system in which arriving 'traffic' can exceed the number of 'servers.'

conjugate *n.* Either of a pair of COMPLEX NUMBERS that are mutually related in that their real parts are identical and the imaginary part of one is the negative of the imaginary part of the other, that is, if

$$z = x + iy,$$

then

$$z^* = x - iy$$

is its conjugate.

conjugate branches Any two BRANCHES of a NETWORK that are related in such a way that a voltage source introduced into one branch produces no current in the other.

Conpernik An iron-nickel alloy. See PERMENORM 5000z.

console *n.* A panel or series of panels containing the controls and indicators for an electronic system.

constant *n.* A quantity that retains a fixed value throughout a given discussion.

Constantan An alloy of 55% copper and 45% nickel used in thermocouples with copper in the temperature range $-169°C$ to $386°C$. Temperature coefficient of electrical resistivity $0.0002/°C$. Spec. gravity 8.9; melting point $1210°C$.

constant-current modulation A form of amplitude modulation in which the plates of the signal amplifier and carrier amplifier (or generator) are fed from the same constant-current source, the signal frequencies thus appearing as modulation of the carrier plate supply and thus of the carrier output.

constant-k filter An IMAGE-PARAMETER FILTER network in which the product of the impedances of the series and shunt branches comprising each section is a constant, that is, if Z_1 is the series impedance and Z_2 the shunt impedance, both functions of frequency, then

$$k^2 = Z_1 Z_2 = R^2$$

where k is a constant and R is the IMAGE IMPEDANCE of the network independent of frequency.

contact *n.* In a switch, relay, etc., any of the conductors that can be moved apart or together to make or break a circuit.

contact bounce CHATTER.

contact microphone A microphone designed to pick up sound as mechanical vibrations of a solid rather than as acoustic waves.

contactor *n.* A relay designed to handle heavy currents, used in electric power circuits.

contact potential The difference of potential that is observed when two dissimilar metals are placed in contact. This effect is due to the difference in WORK FUNCTIONS between the two metals.

contact resistance The resistance measured across a pair of closed contacts.

control grid In an electron tube, the electrode whose potential controls the flow of electrons between the cathode and the plate. In a multigrid tube, this is normally the grid nearest to the cathode.

controlled avalanche A predictable, nondestructive avalanche characteristic designed into a semiconductor device as protection against reverse transients that exceed its ratings.

controlled rectifier A RECTIFIER equipped with a terminal that controls its output current. See IGNITRON, THYRATRON, SILICON CONTROLLED RECTIFIER, etc.

convection *n.* The transfer of mass or energy due to the action of a moving fluid.

convection current See CURRENT.

convergence *n.* In a color kinescope, the condition in which the three electron beams intersect in the plane of the aperture mask and maintain this adjustment as they are deflected.

conversion transducer An electric transducer in which the frequencies of the input and output signals are different.

converter *n.* A FREQUENCY CONVERTER.—**conversion** *n.*

coordinate *n.* Any of the set of numbers that describe the location of a point in space relative to a set of axes.

copper (**Cu;** *Lat.* cuprum) Element 29; Atomic wt. 63.54; Melting pt. 1083°C; Boiling pt. 2300°C; Spec. grav. 8.93 − 8.95; Hardness 2.5 − 3.0; Valence 1 or 2; Spec. heat (20°C solid) 0.092 cal/g; Latent heat of fusion 48.9 cal/g; Latent heat of vaporization 1750 cal/g; Thermal conductivity 0.941 cal/(sec)(cm²)(cm) (°C); Electrical resistivity (20°C) 1.6730×10^{-6} ohm/cm. Used widely in electrical and thermal conductors.

coratio *n.* A parameter used in the description of transformers. The coratio (C) is given by

$$C = (V_2 - V_1)/V_2,$$

where V_2 is the high voltage and V_1 the low voltage.

core *n.* A material of high magnetic permeability placed inside of a COIL to intensify the magnetic field, or, in some cases, to retain magnetism after the magnetizing force has been discontinued.

core loss The loss of energy in the magnetic core of an inductor or transformer due to eddy currents and hysteresis.

core storage In a digital computer, an array of magnetic CORES each of which can be magnetized in either of two polarities and which, therefore, are each able to store one BIT.

corner *n.* A neighborhood or point where a curve makes a sharp or discontinuous change of slope.

corner frequency On a plot of output against frequency, a frequency at which the slope of the curve changes.

corner reflector A reflector consisting of three triangular conducting surfaces. Used in radar for testing and calibration.

corner reflector antenna An antenna consisting of two conductive planes meeting at an angle between 0° and 180° with a driven element consisting of a half-wave dipole included between them.

cosecant *n.* See TRIGONOMETRIC FUNCTIONS.

cosecant squared antenna A radar antenna equipped with a reflector that makes its radiation intensity proportional to the cosecant squared of the direction angle (φ) of (R) the radiation, that is,

$$R(\phi) = k \csc^2 \phi.$$

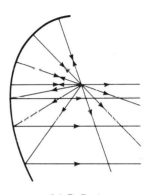

(a) Reflector

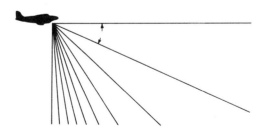

(b) Radiation pattern

cosine *n.* See TRIGONOMETRIC FUNCTIONS.

cotangent *n.* See TRIGONOMETRIC FUNCTIONS.

coulomb *n.* The measure of electric charge, defined as a charge equivalent to that carried by 6.281×10^{18} electrons.

Coulomb's law A law stating that the force (F) between two stationary point charges in free space is given by

$$F = q_1 q_2 / \varepsilon_0 r^2,$$

where q_1, q_2 are the magnitudes of the charges, r is the distance between them, and ε_0 the permittivity of free space. If q_1, q_2 are given appropriate signs, a positive force is one of repulsion and a negative force one of attraction.

counter *n.* **1** A device that (ideally) will switch to the next of *n* discrete states upon the receipt of each input pulse. A counter may be used in conjunction with a READOUT device that indicates which state it is in, or it may be set to deliver an output when it reaches a particular state. **2** A device that counts IONIZING EVENTS, or indicates the average rate at which they occur.

counterpoise *n.* A system of ungrounded metallic conductors used in conjunction with an antenna as a substitute for a ground connection.

counter tube An electron tube that produces a PULSE (ideally) every time it is subjected to an IONIZING EVENT.

countervoltage *n.* The voltage $E = -L\, di/dt$ produced across an inductance (L) as the current (i) through it changes. This voltage acts in opposition to the voltage externally applied.

coupling *n.* A relationship between two systems, networks, etc., such that energy is transferred from one to the other and each influences the behavior of the other. Often the term implies the transfer of an ac component. In any given case it is possible to derive a **coefficient of coupling,** that is, a number between 0 and 1 expressing the closeness of coupling between the systems. See CRITICAL COUPLING, LOOSE COUPLING, OVER-COUPLING.

coupling loop A conductive loop inserted into a waveguide or resonant cavity in order to transfer energy to an external circuit.

covalent bond A chemical bond consisting of a pair of shared electrons, one contributed by each of the bonded atoms.

cps *abbr.* Cycles per second.

crater lamp A glow discharge lamp designed so that the discharge takes place in a depression in its cathode. The lamp acts as a concentrated light source whose brightness is proportional to the current supplied.

creepage *n.* The passage of electricity over the surface of an INSULATOR.

crest factor For a given waveform, the ratio of the peak excursion to the root-mean-square value.

critical coupling The degree of coupling between two resonant systems that results in MAXIMALLY FLAT response around the resonant frequency. Given two systems of the same resonant frequency and QUALITY FACTOR (Q), critical coupling exists between them when $kQ = 1$, where k is the coefficient of coupling.

critical damping In a damped oscillatory system, the minimum amount of damping that renders the response nonoscillatory and causes the system to return to equilibrium in the minimum time without overshoot. In this case the damping (α) is equal to the natural frequency (ω_0). In terms of the CHARACTERISTIC ROOTS this is the case of multiple real roots (α) and therefore the form of the critically damped response is

$$x(t) = K_1 e^{-\alpha t} + K_2 t e^{-\alpha t} + \cdots K_q t^{q-1} e^{-\alpha t}$$

where q is the order of the system and K_1, K_2, . . . K_q are constants determined from initial conditions.

critical inductance In an inductor-input power supply filter, the minimum value of the input inductor needed to insure that the current drawn through the rectifier never goes to zero.

CRO *abbr.* Cathode-ray oscilloscope.

Crookes dark space CATHODE DARK SPACE.

Crookes tube An early type gas discharge tube, filled to a pressure of about 1 mm of mercury and with electrodes located at its ends.

Crosby system The system of FM stereo broadcasting adopted for use in the United States. In this system right channel (R) and left channel (L) are combined to produce an L+R signal and an L−R signal. The L+R signal is transmitted in the normal way and can be demodulated by a monophonic receiver. The L−R signal amplitude-modulates a 38 khz subcarrier which is then suppressed, leaving only the SIDEBANDS. A 19 khz pilot tone, which indicates at the receiver the presence of a stereo transmission, is added. The additional bandwidth remaining to the station is often used as an SCA channel. At the receiver the 19 khz is doubled and used as a reference in demodulating the subcarrier sidebands, and the original L and R are reconstituted from the L + R and L − R signals.

cross-beat *n.* A spurious frequency that arises as a result of cross modulation.

cross coupling Unwanted coupling between circuits.

cross modulation A spurious response that occurs when the carrier of a desired signal intermodulates with the carrier of an undesired signal. This often happens in early stages of radio receivers, particularly when strong signals from local stations drive these stages into nonlinear operation.

cross-over distortion Distortion produced in a class-B push-pull amplifier when the tubes or transistors used are nonlinear in the region of the zero reference point.

crossover network A network that divides an audio signal into two or more frequency bands each of which is fed to an appropriate loudspeaker. A frequency at which equal power is fed to two loudspeakers of the system is called a **crossover frequency.**

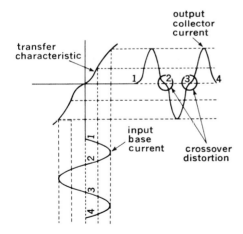

cross product A VECTOR PRODUCT.

cross section A plane section made at right angles to the longitudinal axis of an object, by extension, the probability of a process, particularly a nuclear process, occurring. The cross section is an area (σ) defined by $\sigma = R/IN$, where N is the number of target particles per unit area, I the number of particles incident per second, and R the number of interactions of a particular type.

cross talk Any undesired transfer of signals between two or more different communications channels.

CRT *abbr.* Cathode-ray tube.

cryogenic *adj.* Of or having to do with temperatures approaching ABSOLUTE ZERO.

cryogenics *n.* The branch of physics dealing with temperatures that approach ABSOLUTE ZERO.

cryostat *n.* A refrigeration unit used to produce cryogenic temperatures.

cryotron *n.* A two-PORT device in which the magnetic field caused by the input current controls the superconducting-to-normal transition of the output circuit. A principal advantage of the cryotron is that it dissipates energy only while actually changing states.

cryotronics *n.* The branch of electronics that is concerned with applications of cryogenics.

crystal *n.* Ideally, a solid material in which all of the constituent atoms or molecules occupy predictable positions in a regular, repetitive, geometric pattern. Crystals of this perfection do not exist, all real specimens containing IMPERFECTIONS which strongly influence their physical properties. See PIEZOELECTRIC EFFECT, SEMICONDUCTOR.

crystal detector A DETECTOR that uses a crystal DIODE of one form or another.

crystal diode See DIODE.

crystal microphone A microphone whose active element is a PIEZOELECTRIC crystal that produces a varying voltage in response to variations in sound pressure.

crystal oven A small, temperature-controlled chamber in which a piezoelectric crystal is placed in order to produce maximum stability of its resonant frequency.

C supply A source of voltage used to supply grid bias to one or more electron tubes.

CT *abbr.* Center tap.

CT-cut crystal A quartz crystal cut so that its resonant frequency is below 500 khz.

cub. *abbr.* Cubic.

cubex *n.* An IRON-SILICON ALLOY containing 3% silicon whose crystal structure is cubically oriented.

cu. ft. *abbr.*   **1** Cubic foot. **2** Cubic feet.

cu. in. *abbr.*   **1** Cubic inch. **2** Cubic inches.

Cunife   An alloy containing 20% iron, 20% nickel, and 60% cobalt, and having high coercive force. It is machinable and ductile.

cup core   A CORE that encloses an inductor and acts as a magnetic shield.

curie *n.*   A measure of radioactivity, defined as the quantity of a radioactive nuclide in which there are 3.7×10^{10} disintegrations per second.

Curie point   **1** The temperature above which a FERRO-MAGNETIC material loses its permanent or spontaneous magnetism. **2** The temperature above which a FERRO-ELECTRIC material loses its spontaneous polarization.

curium (Cm)   Element 96; Atomic wt. 243; Chief valence 3; Electronic configuration $5f^7 6d7s^2$. Radioactive with a half-life of 162.5 days; emits alpha particles.

curl *n.*   The vector product of del ∇, the vector differential operator, and another vector **V**, that is, in Cartesian coordinates

$$\text{curl } \mathbf{V} = \nabla \times \mathbf{V} = \mathbf{i}\left(\frac{\partial V_z}{\partial y} - \frac{\partial V_y}{\partial z}\right) + \mathbf{j}\left(\frac{\partial V_x}{\partial z} - \frac{\partial V_z}{\partial x}\right)$$

$$+ \mathbf{k}\left(\frac{\partial V_y}{\partial x} - \frac{\partial V_x}{\partial y}\right).$$

It can be shown that if **V** is the velocity vector of a moving point, then

$$\nabla \times \mathbf{V} = 2\omega,$$

where ω is the angular velocity vector.

current *n.*   The rate at which electric charge is transported across a surface. Analytically $I = dQ/dt$, where I is current, Q charge, and t time. A **conduction current** is due to a flow of charge in a neutral system, as in a metallic conductor or a semiconductor. A **convection current** results from a flow of charges that are not neutralized, as the electrons in a vacuum tube. See DISPLACEMENT CURRENT.

current density   A VECTOR representing current per unit of area in the plane perpendicular to the direction of flow. The direction of the vector is in the direction of positive current flow. Its magnitude is $\lim_{a \to 0} I/a$, where I is current and a area. In general, the current density vector **(I)** varies in magnitude and direction as a function of position and time; in the special case of uniform flow, as in a cylindrical wire, it varies only with time.

current mode logic   Logic circuits in which transistors are fed from constant current sources, keeping them out of saturation and allowing very rapid switching times.

current source   An IDEAL source whose output current waveform is independent of any element or elements connected to its terminals.

cursive *adj.*   Not divided into discrete steps; continuous; analog.

curve tracer   A device that automatically presents the CHARACTERISTICS of an electronic component as a series of curves, often displaying them on an oscilloscope.

cutoff or **cut-off** *n.*   In an electron tube, transistor, etc., a condition in which there is no flow of current in the output circuit, that is, in which no current is drawn from the B-supply; also the bias necessary to produce this condition.

cutoff frequency   A nominal frequency at which a device, network, etc., is considered to cease delivering a useful output. In a theoretical or ideal case this might be a frequency where an abrupt change from zero to infinite attenuation occurs or vice versa. In a practical case it is defined as the frequency at which the ratio of output to input drops by a specified amount, usually 3db.

cutoff voltage   The smallest negative grid voltage that will hold an electron tube in cutoff.

cutout *n.*   A switch, relay, etc., that allows a device to be removed from an electric circuit.

cut set   In a connected electrical network, a set of branches which when open divides the network into two disjunct regions.

CVT *abbr.*   Constant voltage transformer.

CW *abbr.*   Continuous wave.

cy. *abbr.*   **1** Capacity. **2** Cycle. **3** Cycles.

cybernetics *n.*   The science that is concerned with the principles of communication and control, particularly as applied to the operation of machines and the functioning of organisms.

cycle *n.*   A series of sequential changes occurring in a material system in such a way that the system is periodically restored to its initial condition and so that all the sets of changes are equivalent, in particular, the passage of a wave from its equilibrium position to its peak, then to its trough, and back to equilibrium.

cyclic code   A BINARY code in which the combinations representing two successive numbers differ by only one BIT.

cycloconverter *n.*   A device that lowers the frequency of an ac power source to a desired value using controlled rectifiers that are commutated by the ac line.

cyclogram *n.*   On an oscilloscope, the display that results when cyclically related voltages are applied simultaneously to the horizontal and vertical deflection circuits. Compare LISSAJOUS FIGURE.

cycloinverter *n.*   An INVERTER operating in the neighborhood of ten times the output frequency desired, used in conjunction with a CYCLOCONVERTER.

cyclotron *n.*   An ACCELERATOR in which charged particles are raised to high energies by an electric field while confined in circular orbits by a magnetic field perpendicular to the direction of motion. The electric field is varied in synchronism with the motion of the particle, the radius of the orbit increasing as the particle acquires energy.

cyclotron frequency   The angular frequency (ω_c) of a charged particle circulating in a magnetic field, given by

$$\omega_c = qB/m,$$

where q is the charge of the particle, m the mass of the particle, and B is the strength of the magnetic field, all in mks units. At relativistic speeds a correction must be made for the mass of the particle. See RELATIVITY.

D

D *abbr.*   Debye.

DAC *abbr.*   Digital-to-analog converter.

damp *v.*   To act on (an oscillatory system) in such a way as to reduce its amplitude and bring it at last to rest. —**damper** *n.* —**damping** *n.* See DAMPING FACTOR.

damped harmonic motion See HARMONIC MOTION.

damping factor   **1** In a damped oscillatory system, the ratio of the amplitude of one cycle to the following one measured in the same sense or direction. More explicitly, in a system where one of the dynamic variables (u) is given by

$$u(t) = A\, e^{-\alpha t} \cos \omega_d t = \text{real part of } A\, e^{(-\alpha + j\omega)t},$$

where A is the original amplitude, ω_d is the DAMPED RADIAN FREQUENCY, and t is time, the damping factor (δ) is given by

$$\delta = \frac{2\pi\alpha}{\omega_d}.$$

See COMPLEX EXPONENTIAL. **2** A number (d) that represents the ability of an audio amplifier to damp oscillatory motion of a loudspeaker in the absence of a driving signal. Generally $d = Z_1/Z_2$ where Z_1 is the impedance of the speaker and Z_2 the output impedance of the amplifier. This is not equivalent to def. 1.

damping ratio   The ratio of the damping of a system to CRITICAL DAMPING; more explicitly if u, a dynamic variable of the system, is given by

$$u(t) = \text{real part of } A \exp\left(-\alpha + j\sqrt{\alpha^2 - \omega_0^2}\right),$$

then α/ω_0 is the damping ratio. See DAMPING FACTOR.

daraf *n.*   The measure of electrical ELASTANCE, defined as the reciprocal of the FARAD.

dark conduction   In a photoconductive material, residual electrical conduction in total darkness.

dark space   A non-luminous region of a glow discharge tube.

dark-trace tube   A cathode ray tube having a bright (but not necessarily luminescent) screen that absorbs light when bombarded by electrons, and whose trace, therefore, is a dark line. The screen coatings are generally alkali halides.

Darlington connection   A form of COMPOUND CONNECTION in which the COLLECTORS of two or more transistors are connected together and the EMITTER of one is connected to the BASE of the next. Two transistors connected in this way constitute a **Darlington pair.**

D'Arsonval movement   A PERMANENT-MAGNET MOVING-COIL INSTRUMENT.

dashpot *n.* A device in which a fluid is used to damp oscillatory motion, as in a circuit breaker, etc.

data *n.* Information, particularly the material supplied as the input of a computer. *Data* is still construed by many to be a plural form, in computer work, however, *data* is generally construed to be singular.

data processing Any of the operations by which data is sorted, transformed, correlated, classified, etc., so as to be available for further use, especially when these functions are carried out by machines.

db or **dB** DECIBEL(s).

DB *abbr.* Double break.

dBa *abbr.* Decibels above reference noise.

DBB *abbr.* Detector balanced bias.

dbk or **dBk** DECIBEL(S) referred to 1 kilowatt.

dbm or **dBm** DECIBEL(S) referred to 1 milliwatt.

dbp or **dBp** DECIBEL(S) referred to 1 picowatt.

dBrn, dbrn *abbr.* Decibels above reference noise.

dbv or **dBv** DECIBEL(S) referred to 1 volt.

dbw or **dBw** DECIBEL(S) referred to 1 watt.

DC, D.C., dc, d-c, d.c. *abbr.* 1 Direct current. 2 Direct coupled.

d-c amplifier A DIRECT-COUPLED AMPLIFIER.

D.C.C. *abbr.* Double cotton covered (conductor).

DCD *abbr.* Diode-capacitor-diode.

DC dump The disconnection of a computer or a related system from its source of dc power, resulting, in some systems, in the loss of stored information.

dc-restorer A CLAMPER when used to reinsert the dc component of a signal that has been passed through an ac-coupled amplifier.

DCTL DIRECT-COUPLED TRANSISTOR LOGIC.

DDM *abbr.* Difference in depth of modulation.

D.E. *abbr.* Doctor of Engineering.

dead *adj.* 1 Deenergized; not containing or carrying electricity. 2 Inoperative because of a malfunction or fault. 3 Acoustically absorptive.

dead band In a control system, the interval within which the measured variable can vary without causing the system to respond in an effective way.

deadbeat *adj.* Highly damped, as the movement of a meter whose pointer comes to rest quickly and without overshoot.

dead room An ANECHOIC ROOM.

dead short A short circuit of extremely low resistance.

dead time The time interval after the arrival of a pulse, during which a COUNTER is insensitive to the arrival of a second pulse.

deathnium center In a semiconductor crystal, a region in which electron-HOLE recombination goes on more rapidly than elsewhere. Such a region is generally an imperfection or one of the surfaces of the crystal.

DeBroglie wavelength In quantum mechanics, a wavelength (λ) attributed to a particle by virtue of its momentum. In general

$$\lambda = h/mv = \frac{h}{m_0 v} \sqrt{\frac{1 - v^2}{c^2}},$$

where *m* is the observed mass of the particle, m_0 is its rest mass, *v* is its velocity, *c* is the velocity of light, and *h* is Planck's constant.

debug *v.* 1 To free a piece of apparatus of minor malfunctions, especially as one of the stages in its design. 2 To locate and correct the mistakes in a computer PROGRAM.—**debugging** *n.*

dec. *abbr.* Decimeter(s).

decade *n.* The interval between the limits $k \times 10^n$ and $k \times 10^{n+1}$, as, for example, the interval between 30 and 300 or between 100 and 1,000.

decay *v.* To diminish toward some final value.—**decay** *n.*

decay time The time required for some quantity to decay to a specified fraction, generally $1/e$ or about 37 percent, of its initial value.

deci- *combining form.* One tenth (10^{-1}) of a specific quantity or dimension.

decibel A unit used to express ratios of sound or signal power, defined in such a way that *n*, the number of decibels, is given by

$$n = 10 \log_{10} P_2/P_1,$$

where $P_{1,2}$ are the power levels. It follows that if $L_{1,2}$ are levels of current, voltage, or quantities that are their analogs, then

$$n = 20 \log_{10} L_2/L_1.$$

Since decibels represent a ratio it is necessary to establish a reference for 0 decibels in order to indicate an absolute level. For sound this reference is a pressure of 0.0002 microbar. For other references see DBK, DBM, etc.

declinometer *n.* An instrument for measuring the angle between magnetic and geographic meridians.

decode *v.* To transform (information), generally from a form suitable for transmission or data processing to a form suitable for use.—**decoder** *n.*

decouple *v.* To arrange (two or more physical systems) so that coupling between them is either negligible or restricted to the form that is desired.—**decoupling** *n.*

decoupling network A network used to suppress unwanted coupling.

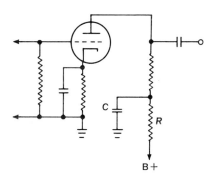

Triode amplifier stage decoupled from the power supply impedance by *R* and *C*.

decrement *n.* **1** A decrease. **2** In a damped oscillating system, the ratio of the amplitudes in two successive periods, measured in the same direction of displacement. The natural logarithm of this ratio is called the **logarithmic decrement.** See DAMPING FACTOR.

decremeter *n.* An instrument for measuring the decrement of a damped electromagnetic wave train.

dee *n.* In a CYCLOTRON, one of the hollow, D-shaped acceleration electrodes.

de-emphasis *n.* The passage of a signal through a network whose frequency response characteristic offsets the characteristic used in PRE-EMPHASIS.

defibrillator *n.* A device for arresting fibrillation by applying electrical impulses to the muscles of the heart.

deflect *v.* To turn aside or cause to turn aside, in particular, to cause a beam of electrons or other charged particles to depart from a straight course by means of an electric or magnetic field.—**deflection** *n.*

deflection electrode An electrode that deflects electrons in a cathode ray tube.

deflection yoke A system of coils through which current is passed, providing a magnetic field that deflects the electron beam of a cathode ray tube.

degauss *v.* **1** To demagnetize. **2** To neutralize the magnetism induced in (a ship) by the earth's magnetic field by passing direct current through a cable that is routed around the hull.—**degausser** *n.*

degenerate semiconductor A semiconductor having a sufficiently large number of electrons in the conduction band that its characteristics approach those of a metal. This condition is often brought about by heavy doping.

degeneration *n.* NEGATIVE FEEDBACK.—**degenerative** *adj.*

degenerative feedback NEGATIVE FEEDBACK.

degree *n.* **1** A measure of angles, equal to $2\pi/360$ radian. **2** A measure of temperature or temperature difference, defined in various ways depending on the system of measurement. The CELSIUS SCALE is most often used in the sciences. **3** In mathematics: **a** Of an algebraic term, the sum of the exponents of its factors. Thus x^3y is of the fourth degree. **b** Of a polynomial or equation, the degree of its highest-degree term. **c** Of a differential equation, the degree of its highest-order derivative.

deka- *combining form* 10 times (a specified unit).

del *n.* The vector differential operator ∇, defined in Cartesian coordinates as

$$\nabla = \mathbf{i}\,\frac{\partial}{\partial x} + \mathbf{j}\,\frac{\partial}{\partial y} + \mathbf{k}\,\frac{\partial}{\partial z}$$

where **i, j, k** are the unit vectors along each of the axes.

delay line A transmission line or its lumped circuit equivalent connected between two points so that a signal will arrive at the second point at a set time after it arrives at the first.

delay time   The amount by which one signal lags another; in particular, in a pulse system, the interval between the time when an input pulse reaches 10 percent of its final value and the time when the output pulse reaches 10 percent of its final value.

delta *n.* (written Δ, δ)   The fourth letter of the Greek alphabet, used symbolically to represent any of various constants, coefficients, etc., of which in electronics the principal ones are: **1** (Δ) An increment. **2** (Δ) PERMITTIVITY. **3** (δ) Deviation. **4** (δ) An angle.

delta function   The generalized function $\delta(t - t_0)$, defined in such a way that

$$\int g(t)\, \delta(t - t_0) = g(t_0),$$

provided that t_0 is in the range of integration. One of its important properties is that

$$\int_{-\infty}^{t} \delta(t) = u(t),$$

where $u(t)$ is a unit STEP FUNCTION at $t = 0$.

Deltamax   An iron-nickel alloy. See PERMENON 5000z.

delta network   A single-mesh network consisting of three branches connected in series.

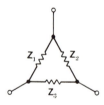

demodulate *v.*   To recover from (a modulated wave) the information imparted by modulation.—**demodulator** *n.*

D.Eng.S. *abbr.*   Doctor of Engineering Science.

density *n.*   **1** The mass of a substance per unit of its volume. **2** By analogy, the number or amount of something per unit of volume or area. See CURRENT DENSITY.

depletion region   In a junction of n-type and p-type semiconductor material, a region where the mobile carriers, electrons and HOLES have recombined leaving a SPACE CHARGE that results from the unneutralized donor and acceptor ions.

derate *v.*   To decrease some rating of (a device or component) in order to compensate for adverse ambient conditions or to increase the reliability of the overall system.

derivative *n.*   The instantaneous rate of change of a FUNCTION with respect to its independent variable, that is, given a function $f(x)$, its derivative $f'(x)$ is defined by

$$f'(x) = \lim_{h \to 0} [f(x + h) - f(x)]/h,$$

provided that this limit exists. Alternative notations for the derivative exist, so that

$$f'(x) = dy/dx = D_x, \text{ and if } y = f(x), y' = f'(x).$$

See PARTIAL DERIVATIVE.

design-center rating   Values of parameters and environmental stresses which should not be exceeded in the normal operation of a bogey electron device.

Destriau effect   Sustained photoemission by certain phosphors when they are embedded in an insulating medium and subjected to an alternating electric field.

destructive readout   In a data-storage system, a readout that erases the stored information.

DET *abbr.*   Detector.

detect *v.*   **1** To DEMODULATE. **2** To indicate the presence of some physical condition, effect, quantity, etc., without necessarily giving a quantitative measurement.—**detector** *n.*—**detection** *n.*

detent *n.*   A mechanical device used to hold a control firmly in each of its several positions.

detune *v.*   To change the frequency of a resonant system away from the desired value, either by changing the product of its inductance and capacitance (or their analogs) or by damping it.

deviation ratio   In a frequency-modulation system, the ratio $\Delta f_c / f_m$, where Δf_c is the maximum shift in the carrier frequency and f_m is the highest modulating frequency.

Dewar flask   A container similar to a thermos bottle, often used to contain cryogenic liquids, as oxygen, nitrogen, etc.

DF *abbr.*   **1** Direction finder. **2** Dissipation factor.

diac *n.*   A BILATERAL TRIGGER DIODE.

diagnostic *n.* A message output by a COMPILER or assembler indicating that a computer program contains a mistake.

diamagnetism *n.* A phenomenon whereby the magnetization induced in certain substances opposes the magnetizing force, in other terms, the negative SUSCEPTIBILITY exhibited by certain substances. The PERMEABILITY of such substances is less than unity.—**diamagnetic** *adj.*

diaphragm *n.* **1** In an acoustic transducer, the element that moves or is moved by the air. **2** A barrier or wall having one or more openings, used to pass a flow of some kind in a selective or controlled manner.

diathermy *n.* The production of heat in living tissue for therapeutic purposes, by means of high frequency electromagnetic waves.

dichroic mirror A type of mirror that reflects light of one color while allowing light of all other colors to pass through it.

dictionary *n.* In digital computer operations, a list of mnemonic code names together with the addresses and/or data to which they refer.

dielectric *n.* A material having a relatively low electrical conductivity; an insulator. The principal properties of a dielectric are its **dielectric constant** (the factor by which the electric field strength in a vacuum exceeds that in the dielectric for the same distribution of charge), its **dielectric loss** (the amount of energy it dissipates as heat when placed in a varying electric field), and its **dielectric strength** (the maximum potential gradient it can stand without breaking down).

dielectric amplifier A parametric amplifier in which the nonlinear element is a ferroelectric capacitor whose capacitance is a function of applied voltage.

difference frequency In a FREQUENCY CONVERTER or other nonlinear transducer, an output frequency equal to the difference of two input frequencies.

differential *n.* Either of a pair of symbols *dy*, *dx* associated with the functional relationship $y = f(x)$ in such a way that

$$dy/dx = f'(x) \quad \text{or} \quad dy = f'(x)dx.$$

Thus it appears that when dy/dx is used as the notation for a derivative it may be treated as a fraction.

While this gives consistent results in some circumstances, it is in general not true.

differential amplifier An amplifier whose output is proportional to the difference between two input signals.

differential equation An equation in which derivatives or differentials of an unknown function appear. The solution of such an equation, if it can be found, is a function, say $g(x)$, that satisfies the equation identically throughout some interval in *x*. The general solution, containing one or more arbitrary constants, represents the set of functions that satisfy the equation. In physical problems, the arbitrary constants are determined from additional conditions that must be satisfied. Some differential equations have singular solutions that are not part of the general solution. Differential equations result from mathematical descriptions of motion and change.

differentiate *v.* **1** To distinguish. **2** To find the derivative of (a function). **3** To deliver an output that is the derivative with respect to time of (the input).—**differentiation** *n.*

differentiator *n.* An electrical network whose output is a close approximation of the derivative with respect to time of its input, that is, whose output is proportional to the rate of change of its input.

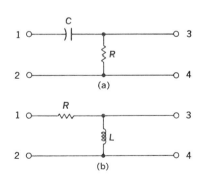

(a) If RC « 1 $V_{34} \cong RC \dfrac{d(V_{12})}{dt}$

(b) If L/R « 1 $V_{34} \cong \dfrac{L}{R} \dfrac{d(V_{12})}{dt}$

diffraction *n.* A phenomenon in which a wave is bent as it passes the edges of an obstacle in such a way that the 'shadow' of the obstacle is reduced in size.

diffuse *v.* To undergo or cause to undergo diffusion. —**diffused** *adj.*—**diffuser** *n.*

diffusion *n.* **1** The intermingling of the molecules (or other particles) comprising two or more substances by means of their random thermal motion. **2** In a semiconductor, the migration of HOLES and electrons under the influence of an electric field.

digit *n.* In a number system based on *m*, that is, in a system in which an arbitrary positive integer is given by

$$q = \sum_{i=0}^{n} c_i \, m^i,$$

any of the set of characters representing the possible values of c_i, that is, the set of integers

$$\{0, 1, 2, \cdots n - 1\}.$$

—**digital** *adj.*

digital computer A computer that processes information in numerical or coded form, rather than in the form of directly measurable quantities. Compare ANALOG COMPUTER.

digital converter A device that puts information into digital form.

digitize *v.* To convert (data) from analog form to digital form.—**digitizer** *n.*

diode *n.* A two-terminal device in which current is a nonlinear function of voltage, and whose characteristic is markedly asymmetrical with respect to the point where voltage equals zero. Diodes are often designed to have other special characteristics as well. See TUNNEL DIODE, VARACTOR, VARISTOR, ZENER DIODE.

diode logic A form of digital circuitry in which logic is performed by varying the biases on diodes so that they turn on and off.

diplexer *n.* A coupling device that allows two radio transmitters to share the same antenna.

dipole *n.* A system composed of a pair of opposite but equal electric charges or magnetic poles separated by a small distance. In a uniform field no force acts on a dipole, but if the axis of the dipole is displaced from the axis of the field by some angle a torque tends to align the dipole with the field. The torque VECTOR **(T)** is given by

$$\mathbf{T} = \mathbf{p} \times \mathbf{E},$$

where **E** is the field strength and **p** the dipole moment. The **dipole moment** is given by

$$\mathbf{p} = q\mathbf{r},$$

where *q* is one of the charges and **r** the distance between the charges.

dipole antenna A straight element, usually about one-half wavelength long, split at its center and connected to a transmission line. Its maximum radiation is in a plane normal to its axis.

dipole layer DEPLETION REGION.

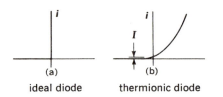

(a) ideal diode   (b) thermionic diode

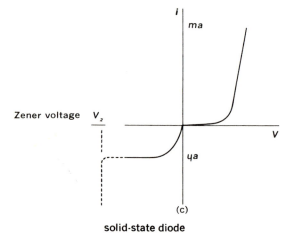

(c) solid-state diode

Diode characteristics

dip soldering A process in which etched circuits and the like are mechanically assembled, coated with flux, and dipped in molten solder. Compare WAVE SOLDERING.

Dirac-h The constant $h/2\pi$, where h is PLANCK'S CONSTANT.

direct-coupled amplifier An amplifier whose response extends to zero frequency, that is, which is capable of amplifying a dc signal.

direct-coupled transistor logic A form of digital circuitry in which logic is performed by switching transistors between cutoff and saturation, the transistors, moreover, being direct-coupled.

direct coupling The coupling of two networks, devices, etc., through a path that allows direct current to pass.

direct current A movement of electric charge through a surface in one direction only.

direct-current amplifier A DIRECT-COUPLED AMPLIFIER.

directional *adj.* Having radiative characteristics that vary with direction.

directional derivative See GRADIENT.

directive gain In an antenna, 4π times the ratio of the radiation intensity in a given direction to the total power that the antenna radiates.

directivity *n.* 1 The directive gain of an antenna in the direction where it reaches its maximum. 2 In a directional coupler, the ratio of the power transferred to the terminals that sense a forward wave to that transferred to the terminals that sense a reverse wave, when only a forward wave is present.

directivity factor In an electroacoustic transducer, the property analogous to the DIRECTIVITY of an antenna, that is, the ratio of the power received or radiated in the direction of maximum response to the total power received or radiated.

directivity index The directivity factor expressed in decibels.

directivity pattern A plot of the response of an electroacoustic transducer as a function of direction.

director *n.* In an antenna array, a parasitic element located a fraction of a wavelength ahead of a driven element or another director in order to increase the directivity and gain of the array.

discharge 1 *n.* The removal of electrical charge from a battery, capacitor, or other storage device. 2 *n.* The passage of an electric current through a gas, accompanied in general by some luminous effect, as a glow or arc. 3 *v.* To give up or lose an electric charge. 4 *v.* To deprive of an electric charge.

discriminator *n.* 1 A circuit that converts variations in frequency or phase into variations in amplitude, often used as a frequency-modulation detector. 2 A device that responds only to pulses having certain special characteristics, as a particular range of amplitudes, durations, repetition rates, etc.

dish *n.* A reflector antenna of the parabolic type.

disk-seal tube An electron tube for use at ultra high frequencies, designed with close electrode spacings and with low electrode lead inductances. Electrode connections are made through disk seals in the envelope.

displacement current A hypothetical current assumed to exist in empty space in the presence of time varying electric fields. Displacement current was postulated by Maxwell to explain the transfer of conduction current through empty space as in the capacitor. In the capacitor the displacement current between the plates equals the conduction current at the terminals, thereby maintaining continuity. In the case of dc or static fields the displacement current is zero. See MAXWELL'S EQUATIONS.

dissector tube An IMAGE DISSECTOR.

dissipate *v.* To lose (energy) in such a way that it becomes unavailable for work. Generally this loss involves a conversion into heat, a part of which contributes to the entropy of the process or system.—**dissapation** *n.*

distortion *n.* Any undesired change in a signal waveform.—**distort** *v.*

distortionless line A transmission line whose PROPAGATION CONSTANT is independent of frequency. This is approached in a practical case by adjusting the line parameters, series inductance (l), shunt capacitance (c), series resistance (r), shunt conductance (g) so that $r/g = l/c$.

distributed amplifier A multi-stage amplifier in which the high-frequency limitation due to the input and output capacitances of the active elements is circum-

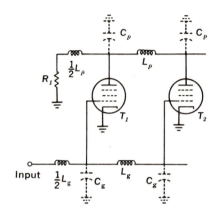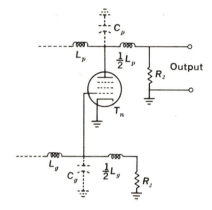

A distributed amplifier $R_1 = \sqrt{\dfrac{L_p}{C_p}}$ $R_2 = \sqrt{\dfrac{L_g}{C_g}}$

vented by making these capacitances the shunt elements of lumped-parameter delay lines. In this way the overall gain is the sum of the gains of the individual stages rather than the product, thus allowing amplification even when the individual gains are less than unity.

dither *n.*   A low-level oscillatory force applied to a servomechanism or other electromechanical transducer to prevent it from sticking in its rest position.

divergence *n.*   The SCALAR PRODUCT of DEL ∇ and another vector **V**. In Cartesian coordinates,

$$\nabla \cdot \mathbf{V} = div\ \mathbf{V} = \frac{\partial V_x}{\partial x} + \frac{\partial V_y}{\partial y} + \frac{\partial V_z}{\partial z}.$$

If at each point in space **V** represents the magnitude and direction of a flow of fluid, *div* **V** gives the rate of decrease of fluid per unit volume. Similarly the divergence can be applied to electric or magnetic flux.

diversity reception   The reception of a radio signal by several antenna and receiver systems whose outputs are selected and combined in order to reduce the effects of fading.

divided-carrier modulation   A form of modulation in which the carrier is divided into two quadrature components each of which is modulated by a different signal. These components are added to form a resultant

that varies in amplitude and phase as a function of the two modulating signals.

divider *n.*   A FREQUENCY DIVIDER or VOLTAGE DIVIDER.

dividing network   A CROSSOVER NETWORK.

DL *abbr.*   Diode logic.

D layer   D REGION.

DM *abbr.*   Double make.

Doherty amplifier   A radio-frequency linear power amplifier consisting of two sections whose inputs and outputs are connected by quarter-wave networks. When the system is driven by a signal voltage of one-half maximum amplitude or less section 1 alone operates; above this level section 2 begins to operate, until at maximum amplitude each section delivers half of the total power to the load.

domain *n.*   **1** In a ferroelectric or ferromagnetic crystal, a region in which the polarization, electric or magnetic, is saturated in a single directon and varies as a function of the temperature only. **2** For a FUNCTION, the set of all possible values of an independent variable.

dominant mode   In a WAVEGUIDE, the transmission MODE having the lowest cutoff frequency. The wave that propagates in this mode is called the **dominant wave.**

dominant wavelength   Of a color sample, the wavelength of light that matches it in chromaticity when mixed with white light.

donor impurity   An element whose presence in minute quantities in a semiconductor creates mobile electrons that can carry current. See N-TYPE SEMICONDUCTOR.

dopant *n.*   An impurity used to DOPE a semiconductor material.

dope *v.*   To change the properties of (a semiconductor material) by adding impurities to it. See N-TYPE SEMI-CONDUCTOR, P-TYPE SEMICONDUCTOR.—**doping** *n.*

Doppler effect   A change in the observed frequency of a train of waves due to motion of the source, the medium, or the observer. For sound waves the observed frequency f_0 is given by

$$f_0 = f_s \cdot \frac{c + v_m - v_0}{c + v_m - v_s},$$

where c is the speed of sound in the medium, v_m is the component of the velocity of the medium along the line between the source and observer, v_s is the same component of the velocity of the source, v_0 is the same component of the velocity of the observer, and f_s is the frequency of the source. For electromagnetic waves.

$$f_0 = f_s \sqrt{\frac{c + v_r}{c - v_r}},$$

where c is the speed of light and v_r is the velocity of the source relative to the observer.

dot cycle   In a communications system, a marking pulse followed by a spacing pulse, each of unit duration.

dot product   A SCALAR PRODUCT.

double-base diode   A UNIJUNCTION TRANSISTOR.

double refraction   A condition in which an anisotropic medium exhibits two indices of refraction for electromagnetic radiation, resulting in the production of two refracted rays for one incident ray. One of these, the **ordinary ray,** remains fixed as the medium is rotated; the other, the **extraordinary ray,** moves with the medium.

double-sideband transmission   The transmission of an amplitude-modulated carrier wave together with both of its sidebands, the carrier often being reduced in strength or even suppressed.

doublet *n.*   A radiating source whose output is equivalent to that of two simple spherical sources located back to back and operating in opposite directions with a phase difference of 180°.

double-V antenna   A DIPOLE ANTENNA modified so that it resembles somewhat a BICONICAL ANTENNA. In comparison with a dipole it has a higher impedance and a broader bandwidth.

down lead or **down-lead**   A length of wire or transmission line connecting a transmitter or receiver to an antenna.

downward modulation   Amplitude modulation in which the carrier has the greatest amplitude when unmodulated.

DP *abbr.*   Double POLE.

D.P.D.T. *abbr.*   Double POLE double THROW (switch).

D.P.S.T. *abbr.*   Double POLE single THROW (switch).

drain *n.*   **1** In a field-effect transistor, the terminal through which current carriers leave. **2** The current delivered to a load by a voltage source.

drain coil   An inductor, having a high reactance at the operating frequency, connected from a broadcast antenna to ground so that lightning and static potentials will leak off.

D region   The part of the ionosphere that extends to about 60 miles above the earth.

dress 1 v.   To arrange (the connecting wires of a circuit) so that undesired coupling is suppressed and the overall appearance is as neat as possible. **2** n. The manner in which the connecting wires of a circuit are arranged.

drift n.   **1** A change, over a period of time, in one or more of the operating parameters of a device or system. **2** The movement of current CARRIERS under the influence of an applied electric field.—**drift** v.

drive 1 n.   EXCITATION. **2** v. To supply excitation to.

driven element   In an antenna array, any element that is connected to the down-lead.

driver n.   In a system, an element whose function is to supply power or excitation to another element, in particular: **1** The penultimate stage of a power amplifier. **2** The electromechanical transducer that feeds a horn loudspeaker.

driving-point immittance   A network function that relates the current and voltage at a pair of terminals at which a source is applied. For a voltage source $v(t)$ the current response is determined by the **driving-point admittance** Y. Thus

$$i(t) = Yv(t).$$

Similarly the **driving-point impedance** Z determines the voltage response to a current source, that is,

$$v(t) = Zi(t).$$

See OPERATIONAL NOTATION.

droop n.   In a rectangular waveform or pulse, a decrease in the height of the nominally flat portion between the leading and trailing edges, usually expressed as a percentage of the peak amplitude.

drop   1 n.   VOLTAGE DROP. **2** v. To develop (a specified difference of potential) between a pair of terminals as the result of a flow of current.

dropout n.   A drastic reduction in the strength of a signal, resulting from a fault in reproduction or transmission.

drum memory   An information-storage device consisting of a rotating cylinder coated with magnetic material. Data is stored in a number of tracks, each served by its own record/playback head.

dry circuit   A circuit in which the voltage is on the order of .03 volt or less and the current .2 ampere or less. In circuits of this type films of oxides, sulfides, etc., on contact surfaces can render them nonconductive.

DSB abbr.   DOUBLE SIDEBAND.

D.S.C. abbr.   Double silk covered.

D-scan   In a radar system, a C-scan in which the bright spots representing targets are vertically elongated, approximately in proportion to their range.

DT abbr.   Double THROW (switch).

DT-cut crystal   A quartz crystal cut so that its resonant frequency is 500 khz or less.

DTL abbr.   Diode-transistor logic.

dual n.   **1** Either of a pair of systems, circuits, etc., that are described by equations of the same form in which

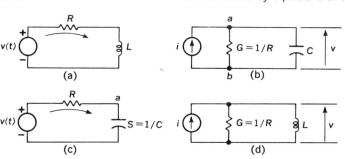

(a) where $v(t) = iR + L\,di/dt$ is the dual of (b)
where $i(t) = vG + C\,dv/dt$; similarly (c)
$\underline{v(t) = iR + S\int_0^t i\,d\tau}$ and (d) $i(t) = vG + L\int_0^t v\,d\tau$
are duals.

the same functional relationships hold provided that the dependent and independent dynamic variables are interchanged between these equations. Thus, a pair of dual circuits might be characterized as follows:

$$v(t) = D_1 i'(t) + D_2 i(t),$$

and

$$i(t) = d_1 v'(t) + d_2 v(t).$$

2 Either of the dynamic variables as $i(t)$, $v(t)$, thus interchanged. **3** Either of a pair of parameters, as D_1, d_1 or D_2, d_2, which are exchanged in transforming these equations one into the other.—**duality** n.

dub v.   To copy (recorded material), often with new material added, as background music, sound effects, etc.

Dumet n.   An alloy of 54% iron and 46% nickel, having high resistivity. Copper-plated, it is used for sealing in leads in electron tubes.

dummy n.   Something used to simulate the electrical characteristics of another device, usually for testing purposes.

duplex adj.   Indicating a communications channel operated in such a way that manual switching is not necessary between transmitting and receiving periods.

duplexer n.   A switching system that allows a single antenna to be used for transmission and reception, while maintaining the necessary isolation between transmitter and receiver.

durchgriff n.   PENETRATION FACTOR.

duty cycle   **1** The fraction of the overall time that a device or system is actually in operation. In a pulsed system, as a radar, the duty cycle is equal to the product of the pulse width and the pulse repetition frequency. **2** DUTY RATIO.

duty factor   DUTY CYCLE.

duty ratio   In a radar system, the ratio of average power output to peak power output.

dyn. abbr.   Dyne(s).

dynamic adj.   Of, concerning, or dependent on conditions or parameters that change, particularly as functions of time.

dynamic analogies   The similarities in form between the differential equations that describe electrical, acoustical, and mechanical systems. These allow acoustical and mechanical systems to be reduced to equivalent electrical networks, which are conceptually simpler than the original systems.

dynamic characteristics   See LOAD CHARACTERISTICS, INCREMENTAL PARAMETERS.

dynamic range   The ratio of the maximum signal level that a system can tolerate to the residual noise level of the system, usually expressed in decibels.

dynamics n.   The branch of physics that deals with the motions of bodies and the effects of forces in producing motion.

dynamo n.   An electrical GENERATOR, in particular, one that delivers direct current as opposed to alternating current.

dynamoelectric amplifier   A generator used as a power amplifier for low frequencies or direct current. The input signal is a varying field excitation; the output is taken from the power windings.

dynamometer n.   **1** An instrument designed to measure forces. **2** An instrument designed to measure mechanical power.

dynamometer movement   An electrical measuring device that operates by means of the torque developed between a fixed coil and a moving coil when currents are passed through them. This torque is proportional to the product of the in phase components of the currents in the two coils. By suitable connections the instrument can be used to measure current, voltage, or power.

dynamotor n.   A rotating electric machine whose armature has two sets of windings, one connected to slip rings for a-c operation, the other to a commutator for d-c operation. In use one set of windings operates as in a motor, the other as in a generator.

dynaquad n.   A germanium PNPN semiconductor device that behaves essentially like a SILICON CONTROLLED RECTIFIER, except that it is triggered by a negative gate signal.

dynatron n.   An inductance-capacitance (LC) sinusoidal oscillator whose active element is a TETRODE operated in its NEGATIVE RESISTANCE region.

dyne n.   In the cgs system, the measure of force, defined as the force which, if applied to a mass of 1 gram, would accelerate it by 1 centimeter per second per second.

dynode *n.* In a photomultiplier or similar electron tube, an electrode designed to emit several secondary electrons each time a single electron strikes it.

dysprosium (Dy) Element 66; Atomic wt. 162.46; Melting pt. 1407°C; Boiling pt. 2600°C; Density 8.54 g/ml; Valence 3; Spec. heat 0.041 cal/(g)(°C); Electrical conductivity (0–20°C) 0.011 $\times$ 10^6 mho/cm.

E

E *symbol* **1** (e) An electron. **2** (e) The charge of an electron. **3** (e) The base of the natural logarithms; 2.78128 . . . See EPSILON. **4** (E or e) Electromotive force. **5** (E or e) Electrode potential. **6** (E) Energy. **7** (E) Electric field vector. **8** (e) Emitter.

earth *n.* *British* GROUND.

ebiconductivity *n.* Conductivity that results from bombardment with electrons. [<e(lectron) b(ombardment) i(nduced *conductivity*]

e.c., ec *abbr.* Enamel-covered.

Eccles-Jordan multivibrator A BISTABLE MULTIVIBRATOR.

echo *n.* A reflected wave that by virtue of its magnitude and delay is perceived as distinct from the directly transmitted wave, in particular, a sound wave thus reflected or a radar signal thus reflected.

echo box A tunable resonant cavity of very high Q used as a synthetic target in the testing of radar systems. The cavity stores energy from the transmitter and gradually returns it to the receiver.

ECM *abbr.* Electronic countermeasures.

ECO *abbr.* Electron-coupled oscillator.

eddy currents Currents induced in the iron core of an electromagnet by variation of the magnetic flux.

Edison effect The thermionic emission of electrons.

EDP *abbr.* Electronic data processing.

E.E. *abbr.* **1** Electrical Engineer. **2** Electrical Engineering.

EEG *abbr.* Electroencephalograph.

effective value For a quantity that varies with time, the ROOT MEAN SQUARE value. See AVERAGE.

efficiency *n.* The ratio of output energy to input energy for a given system; also, by extension, the performance of a system as measured against a theoretically ideal performance—**efficient** *adj.*

EHF, ehf *abbr.* Extremely high frequency.

E-H tee In a rectangular waveguide, a junction composed of an E-plane tee and an H-plane tee that intersects the main waveguide at a common point.

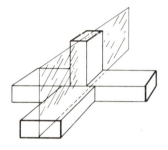

E-H tuner An adjustable E-H tee junction used as an impedance transformer.

EIA *abbr.* Electronics Industries Association.

einsteinium (Es) Element 99; Atomic wt. 254. Radioactive element; emits alpha particles.

elastance *n.* The reciprocal of CAPACITANCE.

elastic collision See COLLISION.

elastivity *n.* The reciprocal of PERMITTIVITY.

electret *n.* A piece of dielectric material that has a permanent electric polarity; the electrostatic analog of a permanent magnet.

electric or **electrical** *adj.* Of, related to, associated with, or operated by electricity. To a great degree *electric* and *electrical* are used interchangeably. Often, however, *electric* is used to designate the physical phenomena of electricity while *electrical* designates things related to electricity. Thus: *electric* charge; *electrical* engineer.—**electrically** *adv.*

electrical axis In a crystal, the axis along which the electrical resistance is a minimum.

electric eye PHOTOELECTRIC CELL.

electric field A condition detectable in the vicinity of an electrically charged body such that forces act on other electric charges in proportion to their magnitudes.

electric field vector A VECTOR representing the force per unit positive charge at a point in an electric field.

electric flux density ELECTRIC INDUCTION.

electric induction A vector **D** representing the charge per unit area induced on a conductor held in an electric field. In an isotropic medium of permittivity (dielectric constant) ε, **D** is given by **D** $=$ ε**E,** where **E** is the ELECTRIC FIELD VECTOR.

electricity *n.* Any of the physical phenomena that involve electric charges and their motions; also, the energy in a system of electric charges.

electric potential The work per unit charge spent in moving a charged body In an electric field from a reference point to a point of interest (*P*). Commonly, the reference point is chosen as infinity. The potential (*V*) is positive if work is done on the charge and negative if work is required of the charge to move in the existing field. Analytically, assuming an electric field of intensity E,

$$V = \int_{\infty}^{p} \mathbf{E} \cdot \mathbf{ds,}$$

where *ds* is a vector element of the path from ∞ to *P*.

electric vector ELECTRIC FIELD VECTOR.

electro- *combining form* Electric: *electroacoustic.*

electroacoustic *adj.* Involving both electricity and sound, as a loudspeaker.

electrobioscopy *n.* The use of electric currents to determine the presence or absence of life.

electrocapillarity *n.* The alteration of the surface tension between two conductive liquids by the passage of an electric current through their interface, resulting in a displacement of the meniscus to an extent that depends on the magnitude and direction of the current.

electrochemical equivalent **1** In electrolysis, the mass of a substance liberated by the passage of one coulomb. **2** The gram-atomic or gram molecular weight of a substance divided by the number of electrons per atom or molecule taking part in an electrolytic reaction. This is the amount of the substance liberated by one FARADAY.

electrochemical series A list of all the elements arranged in order of decreasing ELECTRODE POTENTIALS.

electrochemistry *n.* The phenomena that involve simultaneous or related chemical and electrical changes.—**electrochemical** *adj.*

electrocoagulation *n.* The solidifying of body tissue by means of high-frequency electric currents.

electrode *n.* A terminal through which current flows in passing between the connecting wires of a circuit and another conductor such as an electrolyte, an ionized gas, a semiconductor, etc.

electrode potential The potential of an electrode of an element when it is in equilibrium with a 1 mole per kg aqueous solution of its ions. This potential can only be measured relative to the potentials of other elements under similar conditions. The potential of a hydrogen electrode is arbitrarily defined as zero

electrodynamic *adj.* Having to do with electric charges in motion.

electrodynamics *n.* The part of physics that deals with the connection between electrical, magnetic, and mechanical phenomena.

electrolysis *n.* The chemical decomposition of a conducting substance by the flow of current through it.

electrolyte *n.* A substance that when molten or in liquid solution dissociates into ions that can act as current carriers.

electrolytic capacitor A capacitor in which the dielectric is a film of oxide electrolytically deposited on a plate of aluminum or tantalum. The thinness of the film permits a high capacitance/volume ratio. The oxide acts as a dielectric in one direction only; the device is, therefore, polarized. A **non-polarized electrolytic capacitor** is, in effect, two polarized types in series with their like terminals connected together.

electromagnet *n.* A core of iron or steel which is magnetized by passing a current through a wire coil wound around the core.

electromagnetic constant The speed at which electromagnetic waves propagate in a vacuum, generally designated by *c*. The accepted value for *c* is 299,792.8 km per sec.

electromagnetic wave Any of a class of waves that

propagate as a system of electric and magnetic fields that vary with time.

electromagnetism *n.* Magnetism associated with the flow of an electric current. See AMPÈRE'S LAW.—**electromagnetic** *adj.*

electrometallurgy *n.* The branch of metallurgy dealing with the various processes for the industrial working of metals.

electrometer *n.* An instrument used to measure electric charge, often in terms of the attractive and repulsive forces that operate between electrically charged bodies. In an alternative method the charge is permitted to flow through a high resistance, producing a voltage drop that is measured by what is essentially a vacuum-tube voltmeter.

electrometer tube An electron tube designed so that its control grid conductivity is extremely small.

electromotive force The difference of electrical potential found across the terminals of a source of electrical energy, more precisely, the LIMIT of the potential difference across the terminals of a source as the current between the terminals approaches zero.

electromotive series ELECTROCHEMICAL SERIES.

electrometer tube An electron tube designed so that its control grid conductivity is extremely small.

electromotive force The difference of electrical potential found across the terminals of a source of electrical energy, more precisely, the LIMIT of the potential difference across the terminals of a source as the current between the terminals approaches zero.

electromotive series ELECTROCHEMICAL SERIES.

electromyograph *n.* An electronic instrument for recording contractions and dilations of muscles.

electron *n.* A PARTICLE having a REST MASS of 9.107×10^{-28} g, a charge of magnitude 1.60×10^{-19} coulomb, and a SPIN of one-half unit, that is, of $\hbar/2$ or $h/(4\pi)$, where h is PLANCK'S CONSTANT. Although a negative charge is usually implied, the charge, in general, may be of either polarity. A positive electron is called a **positron,** a negative one a **negatron.**

electronarcosis *n.* The induction of unconsciousness by passage of a weak current through the brain.

electron-coupled oscillator A circuit in which the cathode and two grids of a multigrid tube are connected

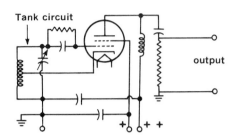

Tank circuit

output

essentially as a triode oscillator, the oscillator being coupled to the load by means of the electron stream that passes on to the plate.

electron gas A system composed of free electrons or quasi-free electrons, as for example, the valence electrons of a metal.

electron gun A device that emits electrons, accelerates them, and focuses them into a beam.

electronic *adj.* Of or having to do with the field of electronics or its principles.

electronic countermeasure A military device or tactic designed to disrupt or interfere with enemy weapons or communications systems that involve electromagnetic radiation.

electronic crowbar An electronic switching device used to divert a fault current from more delicate components until a fuse, circuit breaker, or the like has time to respond.

electronic flash A system in which a capacitor is charged to a fairly high voltage and rapidly discharged through a FLASH TUBE, providing a brilliant pulse of light, as for photography or other applications.

electronics *n.* **1** The branch of science and engineering that deals with the design, manufacture, and utilization of electron devices. **2** The electronic parts of a system or device, as contrasted with the mechanical or other parts.

electronic switch A high-speed switching device using electronic components, in particular, a device that presents two waveforms to the input of an oscilloscope alternately and at a rate sufficiently fast to allow them to be seen simultaneously.

electron lens A system of deflecting electrodes or coils designed to produce on a beam of electrons an effect similar to that produced on light by an ordinary lens.

electron microscope An instrument that projects onto a fluorescent screen the greatly enlarged image of an object held in the path of a sharply focused electron beam, thus permitting visual examination of objects smaller than the wavelengths of visible light.

electron multiplier A device in which a single electron is accelerated in an electric field and allowed to impinge on a special anode (DYNODE) from which it dislodges secondary electrons which, in turn, are treated in the same way. After several stages of this a large pulse emerges for every incident electron.

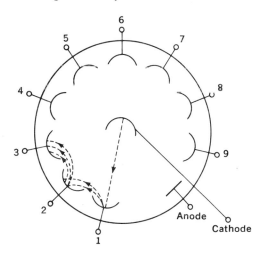

1—9 are dynodes, each at a higher positive potential than the one before, the anode having the highest potential.

electronography n. A kind of printing in which ink is transferred from a printing plate to an impression cylinder by electrostatic means.

electron optics The part of physics that deals with the motions of free electrons in electric and magnetic fields. The laws governing these motions are identical in form to the laws governing light rays that pass between different media.

electron-pair bond A COVALENT BOND.

electron paramagnetic resonance In a metal or semiconductor, PARAMAGNETIC RESONANCE that involves conduction electrons.

electron-positron pair The electron and positron formed in PAIR PRODUCTION.

electron-ray tube An indicator tube whose plate is coated with a material that fluoresces under electron bombardment. It is designed so that either the pattern or magnitude of the glow varies with grid voltage.

electron tube A device in which a current flows due to the passage of electrons through a vacuum or gaseous medium that is isolated from the atmosphere by a gastight envelope.

electron volt A measure of energy, defined as equal to the energy acquired by an electron in passing through a potential difference of 1 volt.

electron wave optics The part of electron optics that deals with diffraction and interference effects.

electron wave tube An electron tube, designed for microwave frequencies, in which a velocity-modulated electron beam passes through an output device, giving up its kinetic energy as radio-frequency energy.

electroosmosis n. The flow of a liquid through a membrane, as a result of the application of an electric current.

electropad n. In an electrocardiograph, the electrode in contact with the skin.

electrophonic effect The auditory stimulation that results when an alternating current of appropriate magnitude and frequency is passed through a person or animal.

electrophoresis n. The movement of electrically charged particles dispersed in a fluid, when under the influence of an electric field.

electrophorus n. A device in which the electric field of an object that has been electrified by friction is used to induce charges in conductors.

electroplate v. To effect the transfer of one metal to another by electrolysis.

electrorefining n. The removal of impurities from a metal by electrolysis.

electroscope n. An instrument for detecting the presence of an electric charge, depending for its action on the mutual repulsion of similarly charged bodies.

electrostatic adj. Of or having to do with static electricity or electrostatics.

electrostatic field　An electric field that does not vary with time, that is, one produced by static charges.

electrostatics *n.*　The part of physics that deals with unneutralized electric charges that are at rest.

electrostatic unit　**1** In the cgs system, the measure of electrostatic charge, defined as a charge which, if concentrated at one point in a vacuum, would repel, with a force of 1 dyne, an equal and like charge placed 1 centimeter away. **2** (*pl.*) A system of electrical units based on the electrostatic unit.

electrostriction *n.*　The development of an elastic strain in certain materials as a result of an applied electric field. This strain is independent of the polarity of the field.

electrothermal *adj.*　Pertaining to the generation of heat by electricity.

electrothermics *n.*　The production of very high temperatures by electrical means, such as resistance heating, induction heating, arc heating, etc.

electrotonus *n.*　The change in the activity of a nerve or muscle when subjected to a steady discharge of electricity.

electroviscous *adj.*　Having increased viscosity due to the application of an electric field.

element *n.*　A substance composed entirely of atoms having a uniform, invariant nuclear charge, none of which can be dissolved by ordinary chemical means.

elementary charge　The magnitude of the charge carried by a single electron or proton, equal to about 1.6 $\times$ 10^{-19} coulomb.

Elf, elf *abbr.*　Extremely low frequency.

ellipse *n.*　A curve consisting of the set of points the sum of whose distances from two other points (the foci) is a constant. In standard form its equation is

$$x^2/a^2 + y^2/b^2 = 1.$$

ellipsoid *n.*　A surface having the property that all its plane sections are ellipses or circles. In standard form its equation is

$$\frac{x^2}{a^2} + \frac{y^2}{b^2} + \frac{z^2}{c^2} = 1.$$

In the special case $a = b = c$ the surface is a sphere.

elliptic(al) polarization　See POLARIZATION.

E.M., EM *abbr.*　Electromotive.

E.M.F., e.m.f., EMF, emf *abbr.*　Electromotive force.

emission tester　An instrument that tests the electron emission from the cathode of an electron tube. Compare TRANSCONDUCTANCE TESTER.

emissivity *n.*　The ratio of the radiant flux emitted by a surface to that emitted by a BLACK BODY at the same temperature and under similar conditions.

emittance *n.*　In a radiating source, the power radiated per unit area of its surface.

emitter *n.*　**1** In a bipolar transistor: **a** The region from which majority carriers (electrons or holes, depending on the type of transistor) originate and are injected into the base, where they either diffuse to the collector or undergo recombination within the base. **b** The terminal that makes contact with the emitter region. The emitter is analogous to the cathode of a tube. **2** In a UNIJUNCTION TRANSISTOR, the region from which holes are injected into the n-type silicon bar between base-one and base-two.

emitter efficiency　In a bipolar transistor, the efficiency of the emitter in injecting carriers into the base in opposition to the leakage current.

emitter follower　A transistor amplifier circuit configuration analogous to a vacuum-tube cathode follower. The circuit is characterized by relatively high input impedance, low output impedance, and a voltage gain of less than unity.

emphasis *n.*　The passage of a signal through a network that in effect accentuates one part of its spectrum. —**emphasize** *v.* —**emphasizer** *n.*

EMR *abbr.*　Electronic Magnetic Resonance.

EMU, E.M.U., emu, e.m.u. *abbr.*　Electromagnetic unit.

encode *v.*　To transform (information) from one representation into another. —**decoder** *n.* See MATRIX.

end effect　A difference between the effective electrical length of an antenna and its physical length caused by capacitance at its ends. The physical length is less than the electrical length by about five percent.

end-fire array　A linear antenna array whose direction (or directions) of maximum response is along its axis.

endodyne reception　AUTODYNE RECEPTION.

endothermic adj.   Absorbing heat.

energy n.   The capability for doing work. Energy is a conserved quantity, that is, it can be neither created nor destroyed. The special theory of relativity relates energy (E) and mass (m) through the equation

$$E = mc^2,$$

where c is the speed of light.

energy level   A stable or quasi-stable state of a physical system in which its energy remains constant for some reasonably long time. See QUANTUM MECHANICS.

ENSI abbr.   Equivalent noise sideband input.

entrainment n.   The shift in frequency of a self-sustaining oscillator when it is excited by a periodic signal whose frequency is close to the natural frequency of the oscillator.

entropy n.   In an isolated physical system, a quantity that expresses the degree to which the energy of the system is unavailable for work. For a thermodynamic system the change in entropy (dS) is given by $dS = dQ/T$, where dQ is the change in heat and T is the absolute temperature. It can be shown for a completely reversible set of changes that $dS = 0$, and for a non-reversible set that $dS > 0$. Thus entropy can never decrease. In an equivalent sense, the entropy of a system can be shown to be greatest when it is in its most probable state. Thus $S = k \ln P$, where P is the probability of the state and k is Boltzmann's constant. In general, the most probable condition is a random distribution of energy throughout the system, hence, in information theory the entropy of a process is a measure of its uncertainty.—**entropic** adj.

envelope n.   **1** The curve defined by the peaks of a sinusoidal wave whose amplitude varies with time, that is, if a modulated wave, M(t) is given by

$$M(t) = A[1 + kV(t)] \cos \omega t,$$

then V(t) is the envelope of the wave. **2** A curve that is tangent to all members of a given family of curves. **3** External housing of an electron tube, light bulb, etc.

envelope delay   The time required for a given point on the envelope of a propagating wave to travel between two points.

epitaxy n.   The growth of a crystal on the surface of a crystal of another substance in such a way that the orientation of the atoms in the original crystal controls the orientation of the atoms in the grown crystal. —**epitaxial** adj.

E-plane   The plane that contains the ELECTRIC FIELD VECTORS of a train of electromagnetic waves.

E-plane bend   In a rectangular waveguide, a bend made so that the longitudinal axis of the waveguide is parallel at all times to the plane of the ELECTRIC FIELD VECTOR of the wave.

E-plane tee- (or **T-**) **junction**   In a rectangular waveguide, a junction in which two arms of the waveguide meet at right angles and their longitudinal axes are parallel to the electric field vectors of their dominant mode of propagation.

EPR abbr.   Electron paramagnetic resonance.

epsilon (written **E, ε**)   The fifth letter of the Greek alphabet, used to represent any of various coefficients, constants, etc., of which in electronics the principal ones are: **1** (ε) Dielectric constant. **2** (ε) Permittivity. **3** (ε) The base of the natural logarithms. See E.

equalizer n.   A NETWORK included in a system in order to compensate for an undesirable frequency response or phase response of some other part of the system. —**equalize** v. —**equalization** n.

equalizing pulse   In television, either of a pair of pulses one occurring just before and the other just after the sync pulse, serving to make each vertical sweep start at the correct time for proper interlace.

equal ripple coupling   See OVERCOUPLING.

equi- combining form   Equal.

equilibrium n.   A condition in which all of the forces or influences acting on a system balance each other out, in other terms, a condition in which at least some of the parameters describing the system are time-invariant. In **stable equilibrium** the potential energy of the system is at a relative minimum, in **unstable equilibrium** it is at a relative maximum.

equilibrium equation   An equation that results from the application of one of Kirchhoff's laws to a network composed of time invariant elements.

equiphase surface   A surface such that all the waves passing through it from a given source are in one of two phases that are 180° apart.

equipotential *adj.* Having a uniform potential at all points.

equivalent 1 *n.* Any of a set of electrical networks or devices that have essentially the same current-voltage relationship between corresponding terminals. **2** *adj.* Being members of a set of equivalents.—**equivalence** *n.*

equivalent circuit A network that is the EQUIVALENT of a more complicated device, used to replace it for purposes of analysis and/or design. Generally, a device does not have a unique equivalent circuit.

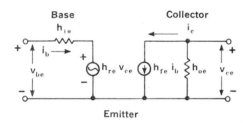

Hybrid equivalent circuit of a common-emitter connected transistor.

erase *v.* To remove written or recorded matter from.

erbium (Er) Element 68; Atomic wt. 167.2; Melting pt. 1497°C; Boiling pt. 2900°C; Density 9.05 g/ml; Valence 3; Spec. heat 0.040 cal/(g)(°C); Electrical conductivity (0−20°C) 0.012 × 10⁶ mho/cm.

E region The layer of the ionosphere between about 60 and 90 miles above the earth.

erg *n.* In the cgs system, the measure of energy, defined as equal to the work done in moving a body 1 centimeter against the force of 1 dyne.

erlang *n.* In telephony, the unit of traffic intensity, defined as the ratio of the number of requests for service during an interval of time to the number of requests that could have been handled if the channel was fully used throughout the interval.

ERP *abbr.* Effective radiated power.

error *n.* **1** In computer work, the deviation of a computed result from a theoretically correct or true result; also the extent or cause of such deviation. **2** A MISTAKE.

error signal See FEEDBACK.

Esaki diode A TUNNEL DIODE.

E scan A radar display in which targets are presented as bright spots on a system of rectangular coordinates, the ordinate indicating elevation and the abscissa range.

ESR *abbr.* Effective signal radiated.

esu, ESU *abbr.* Electrostatic unit.

eta (*written* **H**, **η**) The seventh letter of the Greek alphabet, used symbolically to represent any of various coefficients, constants, etc., of which in electronics the principal ones are: **1** (η) Intrinsic impedance. **2** (η) Efficiency. **3** (η) Surface charge density. **4** (η) Electric susceptibility. **5** (η) Hysteresis.

etched circuit A circuit whose conductors are formed from a sheet of foil bonded to a plastic substrate. Where necessary, isolation between conductors is accomplished by etching away parts of the foil.

Ettinghausen effect The temperature differential that occurs across the edges of a strip of metal held in a magnetic field perpendicular to its surface as a current flows through it longitudinally. The effect is similar to and related to the HALL EFFECT.

Eureka CONSTANTAN.

europium (Eu) Element 63; Atomic wt. 152.0; Melting pt. 826°C; Boiling pt. 1439°C; Density 5.26 g/ml; Spec. heat 0.039 cal/(g)(°C); Electrical conductivity (0−20°C) 0.012 × 10⁶ mho/cm. Used in the red phosphor of color kinescopes.

eutectic *adj.* Referring to an alloy or solid solution that has the lowest possible melting point, usually below that of its components.

EV, ev *abbr.* Electron volt(s).

E vector The ELECTRIC FIELD VECTOR of an electromagnetic wave.

even function See FUNCTION.

E wave *British* A TRANSVERSE MAGNETIC WAVE.

excite *v.* **1** To supply energy to (a system). **2** To supply an input signal to (a network, an active one in particular). **3** To supply current to the field windings of (an electrical generator).—**excitation** *n.* —**exciter** *n.*

exciton *n.* A hole-electron pair in a high energy state bound together in a slab of semiconductor material

The exciton is formed as a result of injecting holes into a neutral crystal. Exciton breakdown, which occurs during recombination, is accompanied by the release of energy in the form of light.

excitron *n.* A mercury-vapor rectifier tube having a pool-type cathode, a single anode, and provision for the maintenance of a continuous cathode spot. The arc is initiated by a solenoid-operated mercury jet, the firing of the main anode being controlled by a grid.

exclusion principle See PAULI EXCLUSION PRINCIPLE.

EXCLUSIVELY-OR gate A circuit having several inputs and one output, with the property that the output becomes energized if one and only one input is energized.

exothermic *adj* Releasing heat.

exp *abbr.* Exponent: used in mathematics to indicate that the expression following is to be considered an exponent of *e*, the base of the natural logarithms.

expanded sweep In a cathode-ray oscilloscope, a sweep whose linear speed (although not its repetition rate) is increased so that a small portion of the displayed waveform fills the entire screen.

expander or **expandor** *n.* An amplifier designed to be complementary to a COMPRESSOR. It tends to increase its gain in the presence of a strong signal and decrease it for weak ones.

exponent *n.* A number indicating the POWER to which an expression is raised, generally written as a superscript at the right of the expression.

exponential function A function of the form

$$f(x) = kb^x$$

where *k*, *b* are constants.

exponential horn A HORN whose rate of flare follows an exponential function.

extinction voltage The lowest voltage that will maintain the discharge in a gas-filled tube.

extraordinary ray See DOUBLE REFRACTION.

extremely high frequency Any wave frequency between 30 and 300 ghz (10^9 hz).

extremely low frequency An electromagnetic wave frequency of 300 hz or less.

F

f *abbr.* Farad(s).

facom *n.* An electronic navigation system that uses phase comparison techniques. If the frequency and phase of a locally generated signal are matched to those of a signal transmitted by a fixed station at one point, the difference in phase at a second point is proportional to the distance between the points.

fade *v.* **1** To gradually change the strength of (a signal). **2** To gradually change in strength, as a signal.

fader *n.* A special mixing network designed so that one signal can be faded in while another is faded out, the output level remaining constant.

Fahnestock clip A spring-type terminal used to facilitate temporary connections.

fail-safe control A control system so designed that although its failure may cause one or more parts of the controlled system to become inoperative, dangerous conditions are avoided.

fall time The time required for a quantity to decrease from 90 percent to 10 percent of its maximum value.

family *n.* **1** In mathematics, a set of functions, curves, etc., that can be generated by varying one or more of the parameters of a general form, thus if

$$G(x,y) = g(x,y, P_1, P_2, \cdots, P_n),$$

a *k*-parameter family of functions is the set of all $G(x,y)$ that result when *k* members of the set $P_1, P_2 \ldots, P_n$ are allowed to take all possible values. **2** A group of chemical elements that have similar or related properties.

fan antenna A DOUBLE-V ANTENNA.

fan-in *n.* The number of inputs that a logic circuit can accommodate.

fan-out *n.* The ability of the output of a logic circuit to drive other such circuits, generally a function of its output impedance.

farad *n.* The measure of capacitance, defined as the capacitance of a condenser that retains 1 coulomb of charge with 1 volt difference of potential.

faraday *n.*   A measure of electric charge equal to 96,-484 coulombs.

Faraday effect   The rotation of the plane of polarization of electromagnetic radiation that passes through a transparent, isotropic medium in the same direction as the lines of force of an applied magnetic field. The angle of rotation (α) is given by

$$\alpha = \omega\, IH,$$

where I is the path length, H the intensity of the magnetic field, and ω a constant of proportionality.

Faraday's law of electromagnetic induction   A law stating that the electromotive force induced in a circuit is proportional to the rate of change of magnetic flux through the circuit, that is,

$$E = -k\, d\phi/dt,$$

where E is electromotive force, ϕ magnetic flux, t time, and k a constant of proportionality.

fatigue *n.*   Structural weakness in a material produced by the repeated application of a cycle of stress.

fault current   A current that flows due to some defect or failure in an electrical system.

FCC *abbr.*   Federal Communications Commission.

F-display *n.*   In a radar system, a display in which the target shows as a bright spot projected onto rectangular coordinates and is located at the origin when the radar is aimed directly at it.

Federal Communications Commission   An agency of the U.S. government that supervises wires, radio, and television communication in the U.S.

feedback *n.*   The return of a part of a system output to the system input, causing, in general, a profound change in the characteristics of the system. If the returned signal is in phase with the input, it is called **positive** or **regenerative feedback;** if it is out of phase it is called **negative** or **degenerative feedback.** Negative feedback improves the stability and linearity of a system at the expense of GAIN, while positive feedback increases gain and speed of response but makes the system less stable and more oscillatory. Consider an amplifier having gain $\mathbf{A_o}$ and distortion $\mathbf{B_o}$, both, in general, complex numbers. It can be shown that via the introduction of a feedback network β the gain $\mathbf{A}_f$ and distortion $\mathbf{B}_f$ of the new closed loop system are made

different from the original open loop parameters. In particular

$$\mathbf{A}_f = \mathbf{A_o}/(1 - \beta\mathbf{A_o}), \quad \mathbf{B}_f = \mathbf{B_o}/(1 - \beta\mathbf{A_o}).$$

Thus, provided that $|1 - \mathbf{A_o}| > 1$, the closed loop system will have lower gain and distortion. It can also be shown that the input impedance is increased and the output impedance decreased, both by the factor $1 - \beta\mathbf{A_o}$. Further, since

$$d\mathbf{A}_f/\mathbf{A}_f = d\mathbf{A_o}/\mathbf{A_o}[1/(1 - \beta\mathbf{A_o})],$$

a change in open loop gain affects closed loop gain but slightly. For a system intended to oscillate sufficient positive feedback must be used so that $1 - \beta\mathbf{A_o} = 0$. See BARKHAUSEN CRITERION; NYQUIST DIAGRAM, BODE DIAGRAM.

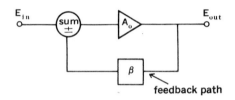

feedback path

feedback control system   A system containing a set of devices that measure its output or outputs against an appropriate set of reference signals, generating a set of error signals which in turn control the system in such a way that its outputs conform to desired performance criteria. The performance, that is, the ratio between output c and reference r, of the generalized system shown is given by

$$c/r = KHG/(1 + KFHG).$$

The system becomes unstable if $1 + KFHG = 0$, that is, if the phase shift around the loop $H\text{-}K\text{-}G\text{-}F$ is zero when $KHFG = 1$. See FEEDBACK.

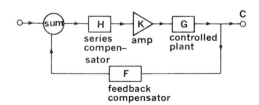

feeder *n.* A conductor or group of conductors connecting different generating or distributing units of a power system.

feedthrough *n.* A conductor designed to pass from one face of a panel to the other.

femto- *combining form* One quadrillionth (10^{-15}) of a specified quantity or dimension.

Fermi-Dirac statistics For a system of particles that satisfy the PAULI EXCLUSION PRINCIPLE, a statistical description of their distribution over a set of energy levels. The **Fermi level** is the energy level for which this distribution function has the value 0.5.

fermium (Fm) Element 100; Atomic wt. 253. All isotopes, which are radioactive, emit beta particles.

ferreed *n.* An electromechanical switch that operates in a latching mode using a combination of bistable magnetic material and metallic contacts.

ferrimagnetism *n.* A type of magnetism that, as the magnetic moments of neighboring ions tend to align antiparallel, appears microscopically similar to ANTIFERROMAGNETISM. The fact that these moments may be of different magnitudes allows a large resultant magnetization that macroscopically resembles FERROMAGNETISM. This property is shown by FERRITES and related compounds.—**ferrimagnet** *n.*—**ferrimagnetic** *adj.*

ferristor *n.* A small SATURABLE REACTOR with two windings, designed to operate at a high carrier frequency.

ferrite *n.* Any of various, often magnetic chemical compounds that contain ferric oxide (Fe_2O_3). The principal ferrites are of the form MFe_2O_4, where M is a divalent metal. Ferrites have high magnetic permeability and high electrical resistivity; thus when used as CORES their eddy-current losses are low.

ferro- *combining form* Iron.

ferroelectricity *n.* A property of certain crystalline materials whereby they exhibit a permanent, spontaneous electric polarization (dipole moment) that is reversible by means of an electric field; the electric analog of ferromagnetism. Materials that show this effect are piezoelectric as well.—**ferroelectric** *adj., n.*

Ferromag I A FERRITE ($BaO \cdot 6Fe_2O_3$) with high coercive force and light weight.

ferromagnetism *n.* A property of certain metals, alloys, and compounds whereby below a certain critical temperature (the CURIE POINT) the magnetic moments of the atoms tend to align, giving rise to a spontaneous, permanent magnetism (dipole moment) that is reversible by means of a magnetic field.—**ferromagnet** *n.*, —**ferromagnetic** *adj.*

ferroresonant circuit A resonant circuit containing a saturable reactor as one of its elements.

ferrospinel *n.* Any of the ferrites that have the crystal structure of spinel, that is, of the mineral $MgAl_2O_4$. These are of the form MFe_2O_4, where M is a divalent metal.

FET *abbr.* (Unipolar) field effect transistor.

fetch *v.* To retrieve information from storage, as in a digital computer.

FF *abbr.* Flip-flop.

fiber optics The branch of engineering and technology concerned with the transmission of light through flexible fibers of various transparent materials. Applications include the transmission of complete images through bundles of fibers.

fidelity *n.* A measure of the degree to which a sound producing system accurately receives and transmits input signals, without distortion.

field *n.* **1** A variable, as electric intensity, temperature, etc., that is a function of the coordinates of the points in a space, that is, a quantity (F) is a field if

$$F = f(X_1, X_2, \cdots, X_n)$$

where $X_1, X_2, \ldots, X_n$ are the coordinates of a point. **2** A region of space in which F, as defined above, has significant values. If F is a scalar the field is a **scalar field;** if F is a vector the field is a **vector field. 3** In television, one of the two complete scanning operations that make up a frame. Where it is necessary to distinguish between defs. 1 and 2, def. 1 may be called a **field quantity.**

field-effect transistor A transistor in which current carriers (HOLES or electrons) are injected at one terminal (the SOURCE) and pass to another (the DRAIN) through a channel of semiconductor material whose resistivity depends mainly on the extent to which it is penetrated by a DEPLETION REGION. The depletion region is produced by surrounding the channel with semiconductor material of the opposite conductivity and reverse-biasing the resulting p-n junction from a control terminal (the gate). The depth of the depletion region depends on the magnitude of the reverse bias. As the reverse-biased junction draws negligible current, the characteristics of the device are similar to those of a vacuum tube.

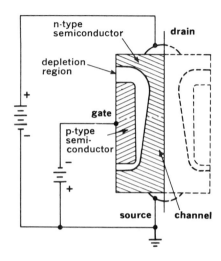

figure of merit A numerical expression of the suitability of a device for a particular application. GAIN, for instance, is a figure of merit for an antenna.

file *n.* In digital computer work, an organized collection of records, usually arranged sequentially, as on magnetic tape.

film resistor A fixed resistor consisting of an insulated form on which a thin film of resistive material is deposited.

filter 1 *n.* An electrical network whose transmission characteristics are a function of frequency. See BAND-PASS, BAND-REJECTION, HIGH-PASS, LOW-PASS, NOTCH, M-DERIVED, CONSTANT-K, etc. **2** *v.* To subject (a signal) to the action of a filter. **3** *n.* A device whose transmission of radiant energy is a function of frequency.

filter capacitor A capacitor that forms part of the low-pass filter used to smooth the output of a power supply.

filter choke An inductor that forms part of the low-pass filter used to smooth the output of a power supply.

filter section Any of various simple networks which may be connected in cascade to form a filter. The simplest is the **half section,** consisting of a series impedance (*Z*) followed by a shunt admittance (*Y*). A **full section** is either a tee-network in which the shunt arm is *Y* and the series arms are *Z*/2 or a pi-network in which the series arm is *Z* and the shunt arms *Y*/2. Full sections, unlike half sections, have equal input and output impedances.

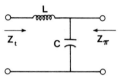

Constant-K low-pass half section

fin *n.* A metal disc or a thin, projecting metal strip attached to a semiconductor to dissipate heat.

fire *v.* To change from a blocked condition, in which negligible current flows, to a saturated condition, in which heavy current flows.

first detector The MIXER or FREQUENCY CONVERTER of a SUPERHETERODYNE RECEIVER.

fishbone antenna An endfire array consisting of a number of paired coplanar elements arranged along a transmission line, resembling somewhat the skeleton of a fish.

fishpaper *n.* A tough, flexible insulating material in sheet form, used to separate coil windings from cores, etc.

five-layer device A semiconductor, as a DIAC, TRIAC, etc., in which there are four p-n junctions.

fixed-point arithmetic In a digital computer, a form of arithmetic in which all numbers are notated by a fixed number of digits, with the decimal point located by implication in a predetermined position.

flashover *n.* A disruptive leakage of current between two parts of a piece of electrical apparatus, or between a piece of the apparatus and ground.

flash tube A gas discharge tube designed to produce brief but very intense flashes of light. The tube is connected to the terminals of a large capacitor that is charged to a high voltage and then ionized, often through a trigger terminal. The capacitor then discharges through the tube, producing a flash.

flat *adj.* Having a SLOPE of zero at all points, as a graph, curve, etc.

F layer One of the layers of ionized air found in the F REGION at varying heights. In the daytime there are normally two such layers, the F_1 **layer** between about 90 and 150 miles above the earth's surface, and the F_2 **layer** which is located somewhat higher. At night the F_2 layer normally exists alone. Radio waves up to about 50 Mhz in frequency are reflected back to earth by the F layers.

Fleming's rule Either of two rules stating that: **1** If the fingers of the right hand curl around a wire so that the thumb points in the direction of current flow, the resulting magnetic field follows the curve of the fingers. (See AMPÈRE'S LAW, AMPÈRE'S RULE.) **2** If the thumb, and the index and middle fingers of the right hand are extended at right angles to each other so that the index finger represents the direction of a magnetic field and the middle finger the direction of current in a wire, the thumb points in the direction of the force on the wire. This is a particularization of the vector relation $F = I(B \times l)$ where **F** is force, **B** the magnetic induction, I the current, and **l** the length of the coil.

Fletcher-Munson effect The variation in the frequency response of the human ear with sound pressure level. See LOUDNESS; LOUDNESS CONTOUR.

flicker effect Small variations in the current passed by a thermionic vacuum tube, caused by random changes in cathode activity. The noise power that results from this is approximately inversely proportional to frequency.

flip chip or **flip circuit** A small semiconductor die all of whose terminals are brought out on one side in the form of solder pads. After passivation of the surface the die is placed with the solder pads in contact with a matching substrate (which may contain thin-film components) and all connections are made simultaneously by application of heat.

flip flip A BISTABLE MULTIVIBRATOR using complementary symmetry transistors. In a circuit of this type both transistors are either saturated or cut off at once.

flip-flop A BISTABLE MULTIVIBRATOR.

float *v.* **1** To be connected to no source of electrical potential: often used with respect to a particular point. **2** To be maintained in a constant state of charge by being connected to a source of constant voltage, as a storage battery.

floating point arithmetic In a digital computer, a form of arithmetic in which each number is represented by several significant digits, with an explicitly placed decimal point, multiplied by the base of the number system raised to a power, as for instance 3.7521×10^{-6}. In computations of this kind the decimal point and exponent are adjusted automatically.

flow chart A schematic representation, similar to a BLOCK DIAGRAM, of a digital computer program.

fluidics *n.* The branch of engineering and technology concerned with the design and production of logic elements, amplifiers, and the like, that depend for their operation on interactions between jets of fluid rather than on electrical phenomena. While slower than electronic logic systems, fluid logic systems can operate in environments that would damage electronic systems.

fluorescence *n.* The process whereby a substance emits electromagnetic radiation as a result of absorbing energy from particle collisions or other electromagnetic radiation, provided that the emission ceases within about 10^{-8} sec after the external energy supply is discontinued.—**fluoresce** *v.* —**fluorescent** *adj.*

fluorescent lamp A tubular, electric discharge lamp in which ultraviolet light from a low pressure mercury arc is converted to visible light after impact upon a coating of phosphors.

fluorine (F) Element 9; Atomic wt. 19.00; Melting pt. $-223°C$; Boiling pt. $-187°C$; Density (20°C) 1.69 g/l; Spec. grav. ($-187°C$ liquid) 1.11; Valence 1; Spec. heat 0.18 cal/(g)(°C).

fluoroscope *n.* An instrument for the direct visual examination of objects put between it and a beam of X-rays. It consists of a fluorescent screen and, generally, an X-ray tube.

flutter *n.* In an audio system, a form of distortion that consists of a rapid frequency modulation of the desired signal. Compare WOW.

flutter echo A train of reflected pulses occuring as a result of a single transmitted pulse.

flux *n.* **1** The number of lines of force that pass through a given surface in a field; analytically flux (ϕ) is given by

$$\phi = K \int_s F dS,$$

where F is a component of the field and dS is an element of area, and K is a constant of proportionality. **2** The flow of mass, energy, fluid, etc., that passes through a surface per unit of time. **3** A material used to retard the oxidation of metals that are being soldered, brazed, or welded.

flux density The FLUX per unit area through a given surface. See MAGNETIC INDUCTION.

flux linkage A quantity (Ψ) considered equal to the number of turns (N) in a coil multiplied by the magnetic flux (ϕ) through the coil, that is

$$\Psi = N\phi.$$

Flux linkage is the DUAL of electric charge.

fluxmeter *n.* An instrument for measuring the total flux linked with a magnetic circuit, generally consisting of a moving-coil galvanometer with negligible restoring torque in the suspension.

flyback *n.* In a sawtooth wave, the shorter of the two periods of time during which the wave passes from one of its extremes to the other.

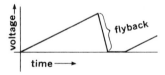

flywheel effect The ability of reactive components, that is, capacitors, inductors, and tuned circuits, to take in energy in pulses and release it in a continuous flow, as does a flywheel.

FM, F.M. *abbr.* Frequency modulation.

f-m stereo A technique of f-m broadcasting in which sufficient information is transmitted to allow a properly equipped receiver to reconstitute both channels of stereo material, while allowing a monophonic receiver to receive a mixture of both channels with no degradation in performance. See CROSBY SYSTEM.

focus *n.* The point at which a system of light rays converge after having passed through an optical system **(real focus)**, or the point at which such rays appear to diverge **(virtual focus)**.

folded-dipole antenna An antenna consisting of a center-fed dipole located close to another parallel dipole, the two being joined at their ends. The impedance is approximately 300 ohms, in contrast to the 70 ohms of a single dipole.

foldover *n.* In television, a type of picture distortion that results from nonlinear horizontal or vertical sweeps. The image appears to overlap itself horizontally or vertically.

foot-candle *n.* A measure of illumination, defined as equal to that produced on a square foot of surface, all points of which are at a distance of 1 foot from a standard candle.

forbidden band A range of energies in which there are no possible electron energy levels.

force *n.* **1** A physical agency that is capable of accelerating or producing elastic strains in material bodies, defined by Newton's second law of motion, $F = ma$, where **F** is force (a vector), m mass (a scalar), and **a** acceleration (a vector). **2** By extension, any causative or operating agency.

forced oscillation The oscillation that results when a system that is capable of free oscillation is subjected to an oscillatory driving force.

forcing function A FUNCTION that describes the characteristics of an external excitation to a given system, as for instance, the time-varying voltage source $v(t) = V_m \cos(\omega t + d)$ connected to the network shown in the figure.

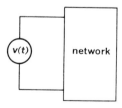

$v(t)$ is the forcing function applied to the network

form *v.* To apply a voltage to (an electrolytic capacitor, semiconductor, or other component) as part of a manufacturing process, in order to cause a desired change in its characteristics.

formant *n.* Any of the resonances in the spectrum of a complex tone which, at least in part, determine its tone color.

format *n.* In computers, a fixed order or arrangement in which input and output are presented, or in which a sequence of internally handled numbers or symbols are interpreted.

form factor 1 For an alternating quantity, the ratio of its peak value to its root-mean-square value. 2 In the expression for the inductance of a coil, a parameter whose value depends on the ratio of the diameter of the coil to the length of the coil.

FORTRAN *n.* A computer-programming language designed mainly for scientific problems.

forward *adj., adv.* In or of the direction in which a non-linear element, as a p-n junction, conducts most easily.

forward point In the current-voltage characteristics of a tunnel diode, the point in the forward bias region where the current is equal to the current at the PEAK POINT.

Foster-Seeley discriminator A PHASE-SHIFT DISCRIMINATOR.

Foster's reactance theorem A theorem stating that the IMPEDANCE (or ADMITTANCE) of a network composed of inductors and capacitors exhibits a number of resonant and antiresonant frequencies that occur in alternation. More explicitly the theorem states that the driving-point IMMITANCE of a two-terminal network composed of a finite number of pure reactances is proportional to an odd, rational function of the COMPLEX VARIABLE $s = j\omega$, and that the POLES and ZEROS of this function lie on the imaginary axis and occur in alternation, with either a pole or a zero at $s = 0$.

Foucault current An EDDY CURRENT.

Fourier series A method of expressing a periodic function as a sum of sine and cosine terms whose frequencies are integral multiples of a fundamental frequency. More explicitly, given a function $f(x)$ defined in the interval $-\pi < x < \pi$ and containing in that interval at most a finite number of finite discontinuities and at most a finite number of maxima and minima, it is possible to express the function in the form

$$f(x) = \tfrac{1}{2}b_0 + \sum_{n=1}^{\infty} b_n \cos nx + \sum_{n=1}^{\infty} a_n \sin nx$$

where

$$a_n = \frac{1}{\pi} \int_{-\pi}^{\pi} f(x) \sin nx dx$$

and

$$b_n = \frac{1}{\pi} \int_{-\pi}^{\pi} f(x) \cos nx dx$$

all for $n > 0$. The term $\tfrac{1}{2}b_0$ is equal to the average value of $f(x)$ over the interval $-\pi$ to π or more explicitly,

$$\tfrac{1}{2}b_0 = \frac{1}{2\pi} \int_{-\pi}^{\pi} f(x)\, dx$$

A Fourier series of a periodic function of time yields the discrete spectral frequency of the waveform. In this case it provides a powerful analytical tool for analysis of periodic processes and the design of engineering systems.

Fourier transform A method of representing a TRANSIENT as a continuous spectrum of frequency components. It might be thought of as a limiting case of the Fourier series in which the period of a function approaches infinity. The result is a continuous rather than a discrete spectrum. Given a non-periodic function $f(t)$ the Fourier transform $g(\omega)$ is given by

$$g(\omega) = \int_{-\infty}^{\infty} f(t)\, e^{-j\omega t}\, dt.$$

An inverse Fourier transform is also defined, in which

$$f(t) = \frac{1}{2\pi} \int_{-\infty}^{\infty} g(\omega)\, e^{j\omega t}\, dt.$$

The functions $f(t)$ and $g(\omega)$ are referred to as a **Fourier transform pair.** There are restrictions on the waveforms that can be transformed by this technique.

four-layer device A PNPN DEVICE.

F.P., f.p, FP, fp *abbr.* Foot-pound.

FPC *abbr.* Federal Power Commission.

fpm, ft/min *abbr.* Feet per minute.

fps, f.p.s., ft/sec *abbr.* Feet per second.

f.p.s. *abbr.* 1 Feet per second. 2 Foot-pound second. 3 Frames per second.

fpsps, ft/s² *abbr.*   Feet per second per second.

Frahm frequency meter   A VIBRATING REED FREQUENCY METER.

frame *n.*   **1** The area occupied by a television picture. In U.S. television this area is scanned completely 30 times per second. **2** One cycle of a train of pulses.

frame grid   In an electron tube, a grid constructed of wires stretched across a rigid, rectangular frame, allowing the use of more and smaller wires than in a conventional grid.

francium (Fr)   Element 87; Atomic wt. 223; Valence 1. Radioactive with a half-life of 21 minutes; emits beta rays.

Franklin antenna   A COLLINEAR ANTENNA.

Fraunhofer region   The region in which the electromagnetic energy from an antenna propagates as though from a point source in the vicinity of the antenna. If the antenna has a well defined APERTURE (D) the Fraunhofer region is considered to begin at a distance $l = 2D^2/\lambda$ from the aperture, where λ is the wavelength of the electromagnetic waves.

free electron   An electron that is not bound to or constrained to remain near a particular atom or nucleus; also, an electron that is subject only to very weak constraints of this type.

free impedance   The impedance measured at the input terminals of a transducer when its output is not loaded.

free-wheel rectifier   A rectifier connected in reverse polarity across an inductive load that is driven from a d-c source, in such a way as to provide a path for the transients that result when the source current is interrupted.

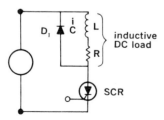

Free-wheel rectifier D_1 bypasses the lagging current through the load, allowing the SCR to turn off.

F region   The region of the ionosphere about 85 to 225 miles above the surface of the earth.

frequency *n.*   The number of occurrences of a periodic phenomenon, as oscillation, per unit time, usually expressed in hertz (cycles per second).

frequency converter   A device whose input and output signal frequencies are different; a CONVERSION TRANSDUCER.

frequency divider   A conversion transducer whose output signal is a proper fraction, usually an integral submultiple, of its input signal frequency. See HARMONIC CONVERSION TRANSDUCER.

frequency frogging   An exchange of carrier frequencies between the input and output of each of a system of repeaters, resulting in greater system stability, less crosstalk, and less need for equalization. In each repeater low-frequency carriers are translated upward while high-frequency carriers are translated downward.

frequency modulation   A form of ANGLE MODULATION in which the frequency of the carrier is made to vary in accordance with the information to be transmitted, that is, given a carrier

$$C(t) = A \cos(\omega_0 t + \phi),$$

and an information function $g(t)$, the signal transmitted is

$$C_m(t) = A \cos[(\omega_0 + g(t) - \Delta\omega)(t) + \phi].$$

When used for radio communications this system has the advantages of greater immunity from noise and other interference, at the cost of increased bandwidth. See CAPTURE EFFECT.

frequency multiplier   A network or transducer whose output signal frequency is a harmonic of its input signal frequency. See HARMONIC CONVERSION TRANSDUCER.

frequency response   A graph showing the ratio of input energy to output energy of a system as a function of frequency.

frequency splitting   In a magnetron, an undesirable condition of operation in which the frequency shifts rapidly back and forth as a result of an alternation of MODES.

frequency synthesizer A device that by heterodyning outputs of several crystal oscillators produces any of a wide choice of discrete frequencies.

Fresnel *n.* A measure of frequency, defined as equal to 10^{12} cycles per second.

Fresnel region The region of space between an antenna and the beginning of the FRAUNHOFER REGION.

Fresnel zone One of the conical zones that exist between the transmission and reception points of electromagnetic radiation because of interference effects between various wavefronts. If D is the direct distance between the points, the outer boundary of the nth Fresnel zone is the surface containing all the paths of length l, such that $l = D + n (\lambda/1)$, where λ is the wavelength of the radiation.

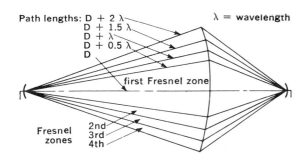

Path lengths: $D + 2\lambda$, $D + 1.5\lambda$, $D + \lambda$, $D + 0.5\lambda$, D
λ = wavelength
first Fresnel zone
Fresnel zones: 2nd, 3rd, 4th

friction *n.* The resistance to relative motion of two bodies, generally greatest, other things being equal, when the bodies are stationary **(static friction)**, becoming somewhat less once they are in motion **(dynamic friction)**.

fringe area The region just beyond the limits of satisfactory radio or television reception.

fringe effect The extension of the FLUX in a field beyond the edges of a gap, as does electric flux at the edges of the plates of a capacitor or magnetic flux at the edges of an air gap in a MAGNETIC CIRCUIT.

frit *v.* To melt and fuse together, as a set of electrical contacts that are subject to repeated discharges.

front end In a radio or television receiver, the stages close to the antenna, generally, in a superheterodyne receiver, including all stages prior to the first intermediate frequency amplifier.

front porch In a COMPOSITE VIDEO SIGNAL, the PEDESTAL between the leading edge of the horizontal blanking pulse and the horizontal sync pulse.

front-to-back (or **-rear**) **ratio** In an antenna, microphone, or other transducer, the ratio of the response in the forward direction (where the response is a maximum) to the response in the opposite direction.

fruit pulse An unsynchronized pulse reply received from a transponder as a result of its being interrogated by a source not associated with the particular responsor

f.s. *abbr.* Foot second.

FSK *abbr.* Frequency-shift keying.

FSM *abbr.* Field strength meter.

FTC *abbr.* Fast time constant.

ft-lb *abbr.* Foot-pound.

fuel cell An electrochemical generator in which the flameless oxidation of a fuel is used to produce electricity.

full section See FILTER SECTION.

full-wave *adj.* Involving both half-cycles of an alternating current.

function *n.* In mathematics: **1** An association between a pair of variables such that for every possible value of the first, the **independent variable,** there corresponds a unique value of the second, the **dependent variable.** More generally, a function may relate any number of independent variables to one dependent variable. If for a given function G,

$$-G(x) = G(-x),$$

G is said to be an **odd function;** if

$$G(x) = G(-x),$$

G is an **even function. 2** The rule that indicates this correspondence: the sine *function;* an exponential *function.* **3** The value of the dependent variable: *y* is a *function* of *x;* Current is a *function* of voltage. ◊ Modern mathematical practice generally insists that a function be single-valued. Consider the relation

$$y = \sqrt{x}.$$

For any value of x, say x_0, there are two possible values of y, namely y_0 and $-y_0$. This is not a satisfactory definition of a function unless the restriction $0 \leq y$ is made, that is, we must exclude negative values of y. Also a function need not be defined for every value of the independent variable. Thus

$$(x+3)^5/(x-1)^2$$

is not defined when its denominator is zero, that is, when $x = 1$.

function generator An electrical network that can be adjusted to make its output voltage (or current) a desired function of time, used in conjunction with analog computers.

fundamental frequency **1** The lowest possible frequency at which a system can execute free oscillation. **2** In a periodic waveform: **a** The greatest common divisor of its component frequencies. **b** The frequency of a sine wave whose period is as long as the period of the waveform.

fuse 1 *n.* A protective device consisting of a strip of metal with a low melting point that dissolves and breaks a circuit if the amperage becomes excessive. **2** *v.* To equip with a fuse.

fused junction See ALLOY JUNCTION.

G

G *symbol* **1** (*g*) Gram. **2** (*g*) Acceleration of gravity. **3** (*G*) Gravitational constant. **4** (*g*) Grid. **5** (*g*) Gate. **6** (*G* or *g*) Electrical conductance.

gadolinium (Gd) Element 64; Atomic wt. 156.9; Melting pt. 1312°C; Boiling pt. 3000°C; Density 7.89 g/ml; Spec. heat 0.071 cal/(g)(°C); Electrical conductivity (0–20°C) 7×10^3 mho/cm.

gain *n.* **1** An increase in signal power in transmission between two points, expressed as a ratio, often in DECIBELS. Often in a system that uses voltage operated devices (that is, devices that draw negligible current), gain is defined as a ratio of two voltages without refer-

ence to the impedances across which these voltages are developed. The **voltage gain** defined in this way is not necessarily equivalent to a power gain. **2** For a directional antenna, the ratio (expressed in decibels) of the maximum response to the response of a standard antenna, often a DIPOLE.

gain-bandwidth product A figure of merit representing the bandwidth over which an amplifying device retains its usefulness, that is, given an amplifier with a midband gain of G, the gain-bandwidth product is $G\,(f_2 - f_1)$, where f_1, f_2 are, respectively, the lower and upper frequencies at which the gain falls to a specified fraction (usually -3db) of G.

gain function A TRANSFER FUNCTION that relates either a pair of voltages or a pair of currents.

gain margin A measure of the stability of a FEEDBACK system, defined as the increase of gain needed to cause oscillation in the system.

gallium (Ga) Element 31; Atomic wt. 69.72; Melting pt. 29.75°C; Boiling pt. 1600°C; Spec. grav. (20°C) 5.91; Valence 2 or 3; Spec. heat 0.079 cal/(g)(°C); Electrical conductivity (0–20°C) $.058 \times 10^6$ mho/cm. Used in semiconductor manufacture, and as a heat-exchange medium in nuclear reactors.

gallium arsenide A chemical compound (GaAs) having useful semiconductor properties which it retains to a temperature of about 400°C.

galvanic series The ELECTROCHEMICAL SERIES.

galvanometer *n.* An instrument for indicating the presence and determining the strength and direction of small electric currents.

gamma (*written* Γ, γ) The third letter of the Greek alphabet, used symbolically to represent any of various coefficients, constants, etc., of which in electronics the principal ones are: **1** (γ) Propagation constant. **2** (Γ) Permeance. **3** (γ) Specific gravity. **4** (γ) Electrical conductivity. **5** (γ) An angle.

gamma rays Quanta of electromagnetic radiation having energies that range between 10^4 and 10^7 electron volts, thus having short wavelengths, high frequencies, and great penetrating power as well. They are emitted by excited atomic nuclei in passing between energy levels.

gang *v.* To arrange mechanical linkages between a set

of similar components, as switches, variable capacitors, etc., so that they can all be operated at once.

garbage *n.* Undecipherable or meaningless sequences of characters produced in computer output or retained within storage.

gas **1** *n.* A state of matter composed of freely mobile particles whose individual volumes are negligible in comparison to the volume through which they are distributed. The salient properties of a gas include low density, high fluidity, and the ability to fill any containing vessel. See IDEAL GAS LAW. **2** *v.* To become GASSY, as a vacuum tube. **3** *v.* To emit gas at an electrode, as a battery or electrochemical cell.

gas constant See IDEAL GAS LAW.

gassy *adj.* Having operating characteristics that are impaired as a result of an excessive amount of gas inside its envelope, as a vacuum tube.

gaston *n.* A modulator that uses a GAS TUBE as a source of NOISE and produces a signal that can be fed to an aircraft communications transmitter and used for jamming.

gas tube An electron tube whose characteristics depend substantially on the gas contained within its envelope.

gate **1** *n.* A network having multiple inputs and a single output which becomes energized when and only when a certain combination of the inputs is energized; a LOGIC CIRCUIT. See AND-GATE, OR-GATE, etc. **2** *n.* A network that transmits (or amplifies) a signal when and only when a certain critical value of a control voltage is reached or exceeded. **3** *v.* To control a current, signal, etc., by or as if by such a network. **4** *n.* The control terminal of a SILICON CONTROLLED RECTIFIER, FIELD EFFECT TRANSISTOR, etc.

gated-beam tube An electron tube that is essentially a PENTODE in which the electrons flowing from cathode to anode are confined in a beam and whose characteristics are such that a relatively small voltage applied to a control terminal produces a sharp cutoff of plate current.

gate turn-off switch A four-layer semiconductor device similar to a silicon controlled rectifier, but with the additional property that a negative pulse applied to its gate will turn it off.

gauge *n.* **1** An instrument or other means for measuring dimension, volume, pressure, etc. **2** The diameter of wires and rods.

gauss *n.* The measure of magnetic induction, defined as equal to 1 maxwell per square centimeter.

Gaussian distribution NORMAL DISTRIBUTION.

Gaussian noise Random electrical noise whose statistical description follows a NORMAL DISTRIBUTION.

gaussmeter *n.* A MAGNETOMETER, especially one calibrated in GAUSS.

Gauss's law A law stating that the total electric flux (F) through a closed surface (s) is equal to the sum (more properly, the integral) of the NORMAL component of the ELECTRIC DISPLACEMENT taken over the surface, which, in turn, is equal to the enclosed charge (q). Analytically

$$F - \int_s \mathbf{D} \cdot \mathbf{ds} = \frac{1}{4\pi} \int_s K\mathbf{E} \cdot \mathbf{ds} = q,$$

where $\mathbf{D}$ is electric displacement, $\mathbf{E}$ is electric intensity, and K the permittivity of the medium within the surface. In differential form,

$$q = \nabla \cdot \mathbf{D}.$$

GAW *abbr.* Gram-atomic weight.

G-display In a radar system, an F-display having the additional property that the spot representing the target undergoes horizontal elongation, as if growing 'wings', as the range to the target decreases.

Geiger-Müller counter or **Geiger counter** A device designed to indicate the presence and intensity of ionizing radiations. The device consists of a gas-filled, cylindrical tube with an anode consisting of a single wire stretched along its axis and a hollow cylindrical cathode surrounding the anode. If a potential exists across the two electrodes, a ray that ionizes the gas will result in a pulse that can be amplified and fed to an indicating device. Proper adjustment of the potential across the tube results in a pulse whose magnitude is independent of the nature and energy of the ray.

general class license A license issuable by the FCC to ham radio operators who are able to send and receive code at the rate of 13 words per minute and who are familiar with general radio theory and practice. Its holder enjoys all authorized amateur privileges except those reserved for higher license classes.

generator *n.* A machine for converting mechanical energy into electrical energy.

geomagnetism *n.* The magnetism of the earth. —**geomagnetic** *adj.*

germanium (Ge) Element 32; Atomic wt. 72.60; Melting pt. 958.5°C; Boiling pt. volatilizes at 2700°C; Spec. grav. (20°C) 5.36; Valence 4; Electron configuration $1s^2 2s^2 2p^6 3s^2 3p^6 3d^{10} 4s^2 4p^2$; Hardness 6–6½Mohs; Latent heat of fusion 114.3 cal/g; Spec. heat (25°C) 0.086 cal/(g)(°C); Thermal conductivity 0.14 cal/(cm)(sec)(°C); Electrical conductivity (0–20°C) 0.022 × 10^6 mho/cm. Having intrinsic semiconductor properties, germanium is used in transistors and crystal diodes.

getter *n.* A highly chemically reactive substance, such as metallic barium, used in electron tubes to trap residual gases.

GeV, GEV *abbr.* Gigaelectron volt.

ghost *n.* An unwanted double image appearing on a television screen, usually caused by multipath signals.

giga- *combining form* A billion (10^9) times a specified quantity or dimension.

gilbert *n.* In the cgs system, the measure of magnetomotive force, defined as equal to 0.7958 AMPERE-TURN.

Gill-Morrell oscillator A type of Barkhausen-Kurz oscillator in which the operating frequency depends on external circuit parameters as well as on electron transit time effects.

gimmick *n.* A capacitor with a value of a few picofarads, improvised by twisting together two insulated wires.

glass electrode A thin wall of specially prepared glass used to separate two solutions, one of known ionic activity and the other unknown. If a standard electrode is dipped into each of the solutions the ionic activity of the unknown can be calculated from the electromotive force developed. While glass electrodes have been principally used in determining hydrogen ion activity, glasses suitable for other ionic measurements have been developed.

glidetone *n.* A device that produces a continuous shift in the frequency of an audio signal, used in electronic music.

glint *n.* In radar, an ECHO that varies in amplitude and apparent point of origin from one pulse to the next because of reflection from a rapidly moving object.

glitch *n.* A stray current or signal, usually one that interferes in some way with the functioning of a system.

glitter *n.* GLINT.

glow discharge One of the modes in which a gas can conduct an electric current, characterized by a glow that surrounds the cathode, a relatively low current density, and a voltage drop that, while considerably higher than the ionization voltage of the gas, is nearly constant for a considerable range of current. In terms of current-voltage characteristics a glow discharge is intermediate between a TOWNSEND DISCHARGE and an ARC DISCHARGE.

glow discharge tube A GAS TUBE whose operating characteristics depend on the maintenance of a GLOW DISCHARGE within it. An important property of such a tube is that the voltage across it remains nearly constant over a considerable range of current.

glow lamp A small glow-discharge tube containing neon or another inert gas at low pressure, used as a light source.

glow tube voltage regulator A glow discharge tube operated in the range of its voltage current characteristics in which the voltage drop across the tube remains fairly constant over a fairly large change of current, often used as a reference in voltage regulator circuits.

gnd *abbr.* Ground.

gobo *n.* **1** A portable shield placed around a microphone to keep out extraneous sounds. **2** A screen for shielding the lens of a television camera from direct rays of light.

gold (Au; *Lat.* aurum**)** Element 79; Atomic wt. 197.2; Melting pt. 1063°C; Boiling pt. 2600°C; Spec. grav. (17.5°C) 19.32; Hardness 2.5–3.0; Valence 1 or 3; Spec. heat 0.031 cal/(g)(°C); Electrical conductivity (0–20°C) 0.42 × 10^6 mho/cm. Used as a cladding material to prevent corrosion of various electronic parts.

goniometer *n.* An instrument used in measuring angles.

go/no go test A test that determines whether or not a given component meets the minimum requirements for a particular application.

Goto pair or **Goto circuit** A series connection of two TUNNEL DIODES such that when one is in the forward conduction region the other is in the reverse tunneling region, used in high-speed switching circuits.

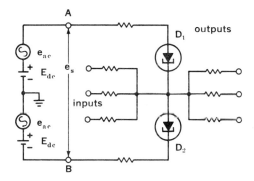

gradient *n.* The gradient of a scalar function (*P*) is a vector in the direction of maximum rate of change of the scalar function and has a magnitude equal to the maximum rate of change. Analytically the gradient is given by

$$\text{grad } P = \frac{\partial P}{\partial s} I_s$$

where *s* is the direction of maximum rate of change. More precisely the gradient is a vector operator ∇ in which

$$\nabla = \frac{\partial}{\partial x} I_x + \frac{\partial}{\partial y} I_y + \frac{\partial}{\partial z} I_z$$

thus

$$\nabla P = \frac{\partial P}{\partial x} I_x + \frac{\partial P}{\partial y} I_y + \frac{\partial P}{\partial z} I_z$$

where I_x, I_y, I_z, are unit vectors in the cartesian co-ordinate system.

gram *n.* The measure of mass in the metric system, defined as equal to one thousandth of a kilogram.

graphechon *n.* A STORAGE TUBE.

graphical analysis Analysis of the behavior of nonlinear devices, such as vacuum tubes and transistors, by reference to curves of their actual operating characteristics rather than by exact calculations.

graphite *n.* A soft, black, chemically inert variety of carbon, used as a lubricant and in making electrical apparatus.

grass *n.* Random noise pulses that produce small disturbances of the baseline of a cathode-ray display, especially in a radar system.

graticule *n.* The network of fine threads or lines in the focal plane of a telescope or other optical instrument, serving to determine the position of an observed object.

gravitation *n.* An effect whereby every particle of matter exerts an attractive force on every other. The force **F** is directed along the line joining the particles, its magnitude being given by

$$F = m_1 m_2 G/s^2,$$

where m_1, m_2 are the masses of the particles, s is the distance between them, and G is a constant of proportionality equal to about 6.670 $\times$ 10^{-8} dyne cm^2/gm^2.

gray body A radiating object whose spectral characteristics are those of a BLACK BODY reduced by a constant factor.

grid *n.* An electrode, usually in the form of a wire screen, mounted between the cathode and anode of an electron tube, that by its potential controls the flow of electrons.

grid base In a TRIODE or PENTODE, the range of GRID voltage between zero and the value needed to drive the tube to CUTOFF.

grid-dip meter An ABSORPTION WAVEMETER consisting of a vacuum-tube oscillator equipped with a meter that reads the average value of its grid current. When the oscillator is coupled to a system that is resonant at its operating frequency energy is drawn from the oscillator, loading it and causing its grid current to decrease or 'dip.'

grid leak A resistor connected in the grid circuit of an electron tube in such a way as to allow charges that

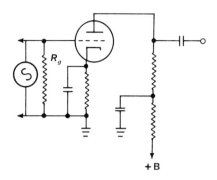

Triode amplifier with R_g the grid leak resistor.

accumulate on the grid (or on the grid capacitor, if any) to leak off.

grille cloth   A loosely woven fabric that is virtually transparent to sound, often stretched across the radiating side of a loudspeaker.

groove *n.*   In phonograph recording, the spiral trace cut into the surface of the record, containing modulation that corresponds to an audio signal.

ground   **1** *n.* In an electric circuit, a point considered, often arbitrarily, to be at zero potential and to which all other potentials in the circuit are referred. In many cases the ground of a circuit is physically connected to the earth. In many electronic applications it is important to distinguish between points that are at ground potential for dc and points that are at ground potential for signals. The output of a well filtered B-supply (with a capacitor output) is clearly not at ground with respect to dc. But since the output capacitor presents a negligible impedance to, say, an rf-signal, such a signal is effectively grounded at this point. **2** *v.* To connect (a terminal wire, etc.) to ground. **3** *v.* To short (a current, signal, etc.) to ground.

grounded *adj.*   Connected to or acting as a GROUND. See COMMON.

ground plane antenna   A vertically polarized antenna that uses the earth as a part of its radiating system. Its important characteristics include nondirectionality and a low angle of radiation.

ground state   The lowest energy level of a quantized system, as, for instance, of a nucleus, atom, or molecule.

ground wave   A radio wave that travels from a transmitting antenna to a receiving antenna along the surface of the ground.

group frequency   The frequency of the variations in the ENVELOPE of a propagating wave.

group velocity   The velocity with which the envelope of a propagating wave travels. The group velocity (u) of a wave is given by

$$u = v - \lambda \frac{dv}{d\lambda},$$

where v is the PHASE VELOCITY and λ the wavelength. Thus in a medium where $v = f(\lambda)$ the group velocity and phase velocity are different.

grown junction   A semiconductor PN JUNCTION produced by varying the concentration and types of DOPANTS used while a semiconductor crystal is grown from a melt.

guard band   An unassigned frequency band between two neighboring channels, left as a safeguard against mutual interference.

guard ring   A ring-shaped electrode intended to limit the extent of an electric field, as, for instance, in elimination of the FRINGE EFFECT at the edges of the plates of a capacitor.

Guillemin line   A type of lumped-parameter delay line used as a pulse-forming network in radar systems.

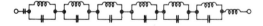

Gunn effect   The development of a fluctuating current in a block of n-type GALLIUM ARSENIDE when a dc voltage exceeding a certain threshold value is applied to contacts on opposite faces. The frequency of the current is in the range of 500 Mhz to 7,000 Mhz and varies with the thickness of the block.

gutta-percha   A coagulated, rubberlike material used in electrical insulation.

gyrator *n.*   A linear, passive element or network that violates the RECIPROCITY THEOREM in that a signal pass-

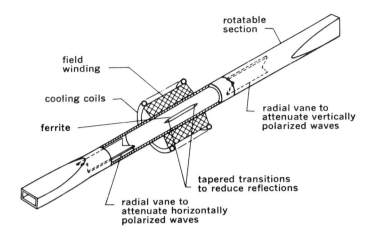

field winding

cooling coils

ferrite

rotatable section

radial vane to attenuate vertically polarized waves

tapered transitions to reduce reflections

radial vane to attenuate horizontally polarized waves

ing through it in one direction undergoes a phase reversal while a signal passing in the other direction is unchanged. These devices are often used in constructing microwave CIRCULATORS and DUPLEXERS.

gyrofrequency n.   CYCLOTRON FREQUENCY.

gyromagnetic adj.   Having to do with the magnetic properties of electric charges in rotation, as electrons moving within atoms.

gyromagnetic effect   The change in the angular momentum of a body as a result of being magnetized, arising as a result of the fact that the magnetic moments of its electrons are associated with their spins or orbital angular momentum.

H

h abbr.   Henry(s).

H symbol   **1** Hydrogen (H). **2** Planck's constant (h). **3** Dirac-h (ℏ). **4** GAUSS (H). **5** Magnetic intensity (H). **6** HENRY (H).

HA abbr.   **1** High altitude. **2** Hour angle.

hafnium (Hf)   Element 72; Atomic wt. 178.6; Melting pt. 1700°C; Boiling pt. above 3200°C; Spec. grav. 13.3;

Valence 4: Spec. heat 0.035 cal/(g)(°C); Electrical conductivity (0–20°C) 0.031 × 10⁶ mho/cm. Used as a filament in lamps, and as a GETTER in vacuum tubes.

half-adder   A digital circuit having one pair of inputs (A,B) and one pair of outputs (C,D) with the condition of the outputs controlled by the inputs according to the following rules:

A-off, B-off = C-off, D-off; A-off, B-on = C-on, D-off;
A-on, B-off = C-on, D-off; A-on, B-on = C-off, D-on.

Two such circuits can be connected to form an ADDER.

half-life or **half-life period**   The period of time during which half the atoms of a radioactive substance will disintegrate.

half-power frequency   Either a high frequency or a low frequency at which the output of an amplifier, network, transducer, etc., falls to one half (−3db) of its maximum or nominal response.

half-power width   In a plane containing the direction of maximum intensity of a radiation LOBE of a directional antenna or an analogous transducer, the angle between the directions at which the intensity of the lobe falls to one half (−3db) of the maximum.

half section   See FILTER SECTION.

half-wave 1 n.   Electrically one half of a wavelength long. **2** adj.   Having to do with one half of a cycle of a wave or alternating current.

Hall effect   A phenomenon characterized by the generation of an electric potential between the edges of a strip of metal through which an electric current is flowing longitudinally, when a magnetic field cuts the plane of the strip at right angles. The magnitude of the effect is measured by the **Hall coefficient** (R_H) which is given by

$$R_H = E_y/J_x H_z,$$

where E_y is the transverse electric field, J_x is the longitudinal current density, and H_z is the vertical magnetic field. The sign of R_H is the same as the charge of the current carriers. Experimental results indicate that the carriers in some solids are positive. See HOLE.

halo *n.*   An unwanted ring of light which develops on the fluorescent screen of a cathode ray tube.

ham *n. informal*   An amateur (but licensed) operator of a radio transmitter.

hangover *n.*   **1** Low-frequency RINGING on the part of an underdamped loudspeaker. **2** TAILING.

hang-up *n.*   **1** An unexplained or nonprogrammed stop in the execution of a computer program, generally the result of a programming mistake or a machine malfunction. **2** The failure of a computer program to operate correctly.

hard *adj.*   **1** Indicating an electron tube that has been evacuated to a high degree. **2** Indicating x-rays of relatively high penetrating power.

hardware *n.*   The mechanical and electronic components of a computer.

harmonic *n.*   A sinusoidal quantity whose frequency is an integral multiple of some fundamental frequency, that is, given a quantity, $x(t)$, where

$$x(t) = A \cos(\omega t + \phi),$$

its harmonics, $h_n(x)$ are of the form

$$h_n(x) = A_n \cos(n \omega t + \phi_n),$$

where n is an integer larger than 1.

harmonic analysis   The process of analyzing a COMPLEX WAVE (or an analogous motion) into its sinusoidal components, that is, in effect, to express a complex wave as a FOURIER SERIES.

harmonic conversion transducer   A CONVERSION TRANS-DUCER in which the frequency of the output signal is an integral multiple or submultiple of the frequency of the input signal.

harmonic distortion   A form of DISTORTION in which, due to the nonlinearity of a network or transducer, the output response to a sinusoidal input contains the sinusoid together with one or more of its HARMONICS.

harmonic mean   See AVERAGE.

harmonic motion   A type of periodic motion characteristically illustrated by a weight oscillating up and down at the end of a spring. The salient characteristic of such motion is that the acceleration of the moving body is proportional to its displacement from its rest position but in the opposite direction. Thus if x is displacement, m the mass of the object, k the coefficient of restoring force, and t time, we have from Newton's law

$$m\frac{d^2x}{dt^2} + kx = 0.$$

It can be shown that

$$x(t) = A \cos\left(\sqrt{\frac{k}{m}}\ t + \phi\right)$$

is the solution of this differential equation. This is **simple harmonic motion.** If there is frictional resistance (R) to motion (assumed proportional to velocity) the resulting differential equation is

$$m\frac{d^2x}{dt^2} + R\frac{dx}{dt} + kx = 0.$$

The solution to this equation has three possible forms (see CHARACTERISTIC EQUATION), but in all of them x approaches a steady state value as t increases. This is referred to as **damped harmonic motion.**

harmonic oscillator   An oscillator whose output is very nearly a sine wave and whose output amplitude and frequency are very nearly constant.

hartley *n.*   A measure of INFORMATION defined as equal to the choice between 10 equally likely alternatives. One hartley can be shown to be equal to $\log_2 10$ BITS.

Hartley oscillator   A sinusoidal oscillator using a three-terminal active element, as a tube, transistor, etc., and a feedback loop containing a parallel LC circuit. The inductor of the LC circuit is provided with an intermediate tap and is connected so as to act like an auto-transformer in matching the input and output impedances of the active device.

hash *n.* **1** Electrical noise that results from rapid opening and closing of contacts. **2** GRASS.

Hay bridge An ac BRIDGE by means of which inductance can be measured in terms of resistance, capacitance, and frequency. This bridge is most suitable for inductors that have a high value of *Q* (QUALITY FACTOR).

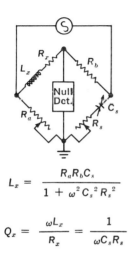

$$L_x = \frac{R_a R_b C_s}{1 + \omega^2 C_s^2 R_s^2}$$

$$Q_x = \frac{\omega L_x}{R_x} = \frac{1}{\omega C_s R_s}$$

hazard *n.* The probability of failure or malfunction in a device or system.

H-display In a radar system, a B-DISPLAY altered so that angle of elevation is included. A target shows as two closely spaced dots that approximate a line, whose slope is proportional to the sine of the angle of elevation.

head *n.* A MAGNETIC HEAD.

headphone *n.* A telephone receiver, usually attached to a flexible band worn about the head.

headset *n.* A pair of headphones.

heater *n.* An element in an electron tube that heats the cathode to the temperature of emission.

heat shunt A HEAT SINK placed in contact with a lead of a delicate component in order to provide protection during soldering.

heat sink A mass of metal placed in physical contact with a device in order to conduct away the heat it produces, disposing of the heat finally by convection or radiation.

Heaviside-Campbell mutual inductance bridge A variation of the HEAVISIDE MUTUAL INDUCTANCE BRIDGE in which one of the inductive arms contains a separate inductor that is included while one set of measurements is made and shorted while a second set is made.

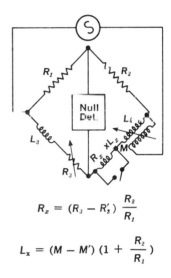

$$R_x = (R_3 - R_3')\,\frac{R_2}{R_1}$$

$$L_x = (M - M')\left(1 + \frac{R_2}{R_1}\right)$$

Heaviside layer A region of the ionosphere about 60 miles above the earth that reflects radio waves of relatively low frequency back to the earth.

Heaviside mutual inductance bridge An ac BRIDGE used to compare mutual inductances and also self inductances.

hecto- *combining form* A hundred (10^2) times a specified quantity or dimension.

heelpiece *n.* The bar of iron connecting the soft iron cores in an electromagnet.

Heising modulation CONSTANT-CURRENT MODULATION.

helical antenna An antenna whose driven element is a helix supported above a ground plane. If the circumference of the helix is of the order of one wavelength the radiation is confined in the main to a lobe running along the axis of the helix. In addition, the radiation is circularly polarized.

helical potentiometer   A potentiometer, normally of the high-precision type, in which the resistive element takes the form of a helix. Several turns of the control knob are needed to move the contact arm through its full range.

helitron *n.*   A BACKWARD-WAVE (oscillator) TUBE that uses electrostatic rather than magnetic focusing, achieving thus a substantial reduction in total weight.

helium (He)   Element 2; Atomic wt. 4.003; Melting pt. below $-272.2°C$ (26 atm); Boiling pt. $-268.9°C$; Density 0.177 g/l; Valence 0; Spec. heat 1.25 cal/(g)(°C).

helix *n.*   A curve traced on the surface of a cylinder or cone in such a way that each element of the surface is cut at the same angle; the curve followed by the thread of a screw or bolt.

henry *n.*   The measure of inductance, defined as the inductance of a circuit in which a counter electromotive force of 1 volt is generated when the current is changing at the rate of 1 ampere per second.

heptode *n.*   A seven-electrode electron tube containing an anode, cathode, and five grids.

hermaphroditic connector   Either of a pair of coaxial connectors whose mating faces are alike.

herringbone pattern   In television, a form of interference sometimes seen on a receiver as a closely spaced band of V-shaped or S-shaped lines.

hertz *n.*   The measure of frequency, defined as equal to 1 cycle per second.

Hertz antenna   A DIPOLE ANTENNA.

heterodyne *v.*   To mix (ac signals of different frequencies) in a nonlinear device. In general, the output voltage (E_{out}) of such a device is given by

$$E_{out} = \sum_{n=1}^{\infty} A_n (E_{in})^n,$$

where E_{in} is the sum of all the input voltages. For most practical applications E_{in} consists of but two frequencies. In this case, assuming an appropriate choice of the nonlinear device, it is reasonable to ignore the terms where $n > 2$ in the expression for E_{out}. Thus, practically, the output consists of the two original fre-

quencies together with their sum and difference frequencies. See FREQUENCY CONVERTER, INTERMODULATION, SUPERHETERODYNE.

hexode *n.*   A six-electrode tube containing an anode, cathode, and four grids.

HF, hf *abbr.*   High frequency.

high band   The band of television frequencies from 174 Mhz to 216 Mhz, including channels 7–13.

high fidelity   The reception and/or reproduction of a signal with a minimum of distortion.

high frequency   A radio frequency in the band between 3,000 and 30,000 khz.

high-pass filter   A FILTER that ideally passes all frequencies above some nonzero CUTOFF FREQUENCY with negligible attenuation while providing virtually infinite attenuation at frequencies below that.

high tension   High voltage.

hill-and-dale recording   A recording in which the modulation of the groove is made perpendicular to the surface of the disc.

Hiperco 35   An alloy of 64% iron, 35% cobalt, and 1% chromium, having high permeability and a high saturation flux density.

Hipernik V   An alloy containing 50% nickel and 50% iron, grain oriented. See PERMENORM 5000z.

Hipersil *n.*   An IRON-SILICON ALLOY containing 3.25% silicon; used in transformer cores.

H-network   A two-port network containing five impedance branches arranged in the form of a letter H lying on its side.

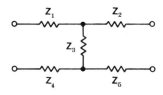

hog horn   A radar horn in which a flared out waveguide provides a parabolic surface for reflection of the electromagnetic waves.

hold *v.*   To retain the information contained in one storage location of a computer after copying the information into another storage location, as opposed to clearing or erasing the information.

holding current 1 The minimum current that must be passed through the coil of a RELAY in order to keep it activated. 2 The minimum current that must pass through a device such as a silicon controlled rectifier, thyratron, neon glow tube, etc., to maintain it in a conducting condition.

hole *n.* In a solid, a vacant energy level near the top of a BAND. The description of such a band is generally more convenient if it is considered to contain a few holes rather than being full of electrons. Experimental results confirm the notion that holes are mobile and that they behave very much like positively charged particles. See HALL EFFECT.

holmium (Ho) Element 67; Atomic wt. 164.94; Melting pt. 1461°C; Boiling pt. 2600°C, Density 8.80 g/ml; Valence 3; Spec. heat 0.039 cal/(g)(°C); Electrical conductivity (0–20°C) 0.011 × 10⁶ mho/cm.

homodyne detector A PRODUCT DEMODULATOR designed to operate with an amplitude modulated signal.

honeycomb winding A method of winding a coil with criss-cross turns to minimize distributed capacitance.

hookup *n.* 1 A circuit diagram for a piece of electronic apparatus. 2 The circuit itself.

hookup wire A soft-drawn copper wire, either solid or stranded, of about 18–20 gauge, used in assembling electronic equipment. In most cases this wire is tinned and insulated.

hop *n.* A single passage of a radio wave to the ionosphere, where it is reflected, and back to earth again. The number of hops taken is referred to as the **order of reflection.**

horizontal sync pulse In television, any of the series of pulses that serve to synchronize the horizontal scanning at the receiver with that at the transmitter.

horn *n.* 1 A flared tube that acts essentially as a tapered acoustic transmission line. It is driven at its smaller end by a loudspeaker having a relatively high mechanical impedance, and feeds into the surrounding air, which has a relatively low impedance. The horn serves to match these impedances, and, in addition, determines to a large extent the direction of radiation. 2 A HORN ANTENNA.

horn antenna A microwave antenna consisting of a flare at the end of a waveguide. Its operation is somewhat analogous to that of an acoustical horn, in that it matches the impedance of the waveguide to the impedance of free space and simultaneously provides a desired radiation pattern.

horsepower *n.* A measure of power, defined as equal to 33,000 foot pounds per minute, or 746 WATTS.

hot *adj.* 1 At a high temperature. 2 Strongly radioactive. 3 Excited to a relatively high energy level. 4 Connected to a source of voltage or current; energized.

hot-cathode tube An electron tube whose operation depends on thermionic emission from its cathode.

hot-wire instrument A measuring device or transducer whose operation depends either on the expansion of a wire due to its being heated by an electric current or on the change in electrical resistance on the part of a wire that is heated or cooled.

housekeeping *n.* The setting up of storage area, constants, etc., prior to the execution of a computer program, performed either by the COMPILER or by the programmer himself.

hp *abbr.* Horsepower.

H.P., h.p. *abbr.* High power.

h-parameters HYBRID PARAMETERS.

HPF *abbr.* Highest probable frequency.

H-plane The plane that contains the magnetic field vectors (H-VECTORS) of a train of electromagnetic waves.

H-plane bend In a rectangular waveguide, a bend made so that the longitudinal axis of the waveguide is parallel at all times to the plane of the H-VECTOR of the wave.

H-plane tee- (or T-) junction In a rectangular waveguide, a junction in which two arms of the waveguide meet at right angles and their longitudinal axes are parallel to the H-VECTORS of their dominant mode of propagation.

H.T. *British abbr.* High tension.

hue *n.* A subjective attribute of a color that corresponds to the dominant or complementary wavelength of the light that composes it. Hue determines whether a color is perceived as red, yellow, or blue, etc. White, black, and gray are not considered hues.

hum *n.* **1** An extraneous low-frequency signal, particularly one that is at the power-line frequency or one of its multiples. **2** *v.* To produce a hum.

hum bar In television, a bar extending across the picture, arising as a result of a power-line hum that leaks into the video signal.

hunting *n.* In a feedback control system, an instability that results in the controlled variable oscillating about the desired value rather than settling down.

H.V., h.v., Hv, hv *abbr.* High voltage.

H-vector A vector representing the magnetic field component of an ELECTROMAGNETIC WAVE.

H wave *British* A TRANSVERSE ELECTRIC WAVE.

hy *abbr.* Henry(s).

hybrid circuit An electronic circuit in which two different types of components are used to perform essentially the same function, as, for instance, a circuit containing both tubes and transistors.

hybrid junction A network having four pairs of terminals and the property that a signal entering at one pair of terminals is divided between the two adjacent terminal pairs without appearing at the opposite terminal pair. Often such a device consists of a junction of four waveguides.

hybrid parameters A set of four parameters (h) derived from a four-terminal EQUIVALENT CIRCUIT of a transistor which conveniently specify the transistor's small-signal characteristics. With the output shorted, that is,

$$v_2 = 0, h_{11} = h_i = v_1/i_1 = \text{input impedance,}$$

and

$$h_{21} = h_f = i_2/i_1 = \text{forward transfer current ratio.}$$

With the input open-circuited, that is

$$i_1 = 0, h_{12} = h_r = v_1/v_2 = \text{reverse transfer voltage ratio,}$$

and

$$h_{22} = h_o = i_2/v_2 = \text{output admittance.}$$

A second subscript is used to denote the transistor con-

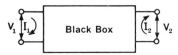

nection; thus h_{fe} is the forward transfer current ratio in the common-emitter mode. These parameters are useful in analyzing cascaded transistor circuits.

hybrid ring A HYBRID JUNCTION composed of a reentrant coaxial line or waveguide whose length is adjusted so as to sustain standing waves, and to which four side branches are connected at appropriate electrical intervals.

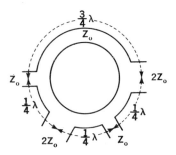

hybrid tee or **hybrid T** A HYBRID JUNCTION for microwaves, consisting of an E-H tee with internal matching elements. A wave entering one arm is not at all reflected provided that the other three arms are match-terminated.

hydrogen (H) Element 1; Atomic wt. 1.008; Melting pt. $-259.14°C$; Boiling pt. $-252.7°C$; Density 0.08988 g/l; Spec. grav. ($-252°C$ liquid) 0.070; Valence 1; Critical temp. $-239.9°C$; Critical pressure 12.8 atm; Critical density 0.0301 g/cm^2; Heat of fusion ($-259.2°C$) 14.0 cal/(g)(°C); Heat of vaporization ($-252.8°C$) 107 cal/g; Specific heat at constant pressure ($25°C$ gas) 3.42 cal/(g)(°C), ($-256°C$ liquid) 1.93 cal/(g)(°C), ($-259.8°C$ solid) 0.63 cal/(g)(°C); Thermal conductivity ($25°C$) 0.000444 cal/(cm)(cm²)(sec)(°C); Viscosity ($25°C$) 0.00892 centipoise. Hydrogen has a high thermal conductivity, and is used as an atmosphere for high-temperature welding, cutting, and melting of materials.

hydrogen electrode An electrode whose potential is that of hydrogen, generally a form of GLASS ELECTRODE.

hydrometer *n.* A device for determining the specific gravity or density of a liquid by measuring the level to which a float of known density sinks while floating in the liquid.

hydrophone *n.* An electroacoustic transducer designed to respond to underwater sound waves and deliver essentially equivalent electric currents.

hygroscopic *adj.* Able to absorb and retain moisture from the atmosphere.

Hymn 88 An alloy containing 79% nickel, 4% molybdenum and 17% iron and having high permeability and resistivity.

hyperbola *n.* A curve consisting of the set of points the difference of whose distances from two fixed points is a constant. In standard form its equation is

$$x^2/a^2 - y^2/b^2 = 1.$$

In parametric form

$$x = \cosh\theta, y = \sinh\theta.$$

—hyperbolic *adj.*

hyperbolic exponential horn Any of a family of acoustical HORNS whose cross-sectional area (A) is given by

$$A = A_t[\cosh(x/x_0) + T\sinh(x/x_0)]^2,$$

where A_t is the area of the throat, x_0 is the reference axial length, and T is a parameter that determines the member of the family. If $T = 1$ the horn is an EXPONENTIAL HORN.

hyperbolic functions A set of TRANSCENDENTAL FUNCTIONS having properties similar to those of the TRIGONOMETRIC FUNCTIONS. The relations between these functions and a hyperbola are analogous to the relations between the trigonometric functions and a circle. These functions, which can arise in the descriptions of networks, are the **hyperbolic sine** (sinh), **hyperbolic cosine** (cosh), **hyperbolic tangent** (tanh), **hyperbolic cotangent** (coth), **hyperbolic secant** (sech), and **hyperbolic cosecant** (csch). Their definitions are:

$$\sinh z = \frac{(e^z - e^{-z})}{2} = z + \frac{z^3}{3!} + \frac{z^5}{5!}\cdots;$$

$$\cosh z = \frac{(e^z + e^{-z})}{2} = 1 + \frac{z^2}{2!} + \frac{z^4}{4!}\cdots;$$

$\tan z = \sinh z/\cosh z; \coth z = 1/\tanh z; \text{sech } z = 1/\cosh z; \text{csch } z = 1/\sinh z.$ Furthermore, it can be shown that $\sinh iz = i\sin z$ and $\cosh iz = \cos z$.

hypersonic wave An acoustic wave having a frequency of 10^9 hz or above.

Hypex horn A HYPERBOLIC EXPONENTIAL horn whose parameters are adjusted so that its response characteristic reaches a peak at some frequency approaching its CUTOFF FREQUENCY.

hysteresis *n.* The effect of the previous history of a system on its response to an external force or influence. Usually the term refers to a magnetic effect, but it may be applied equally well to effects found in dielectrics, switching systems, etc. Consider a functional relationship between a pair of variables, (for discussion let B, magnetic induction, be the independent variable and H, magnetic field strength, be the dependent variable) such that if both are initially zero and B increases, H increases in a characteristic way until SATURATION is reached (curve $t_0 - t_1$ in illustration). If at this point B starts to decrease H decreases also but through a new set of values, finally reaching saturation in the reverse direction (curve $t_1 - t_2$). Should B increase again H increases through yet a third set of values toward the first saturated condition (curve $t_2 - t_1$). Further variation of B between these extremes results in repeated tracing out of the curve $t_2 - t_1, t_1 - t_2$, the **hysteresis loop.** If the variation of H involves a transfer of energy there is a loss proportional to the area within the loop. This is called **hysteresis loss.**

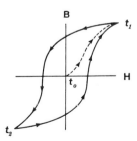

hz *abbr.* Hertz.

I

I *symbol* **1** (*I*) Moment of inertia. **2** (*I*) Current. **3** (*I*, *i*) Instantaneous current. **4** (*i*) Square root of −1. **5** (*i*) Unit vector on the x-axis. **6** (*I*) Acoustic intensity.

IAGC *abbr.*   Instantaneous automatic gain control.

IC *abbr.*   Integrated circuit.

iconoscope *n.*   A television camera tube containing a photoemissive mosiac screen upon which the optical image is focused and which is scanned by an electron beam. The electron beam restores to each photoemissive element the charge removed by the incident light. Thus the magnitude of each pulse of current is proportional to the brightness at that point of the scan.

i.c.w., ICW *abbr.*   Interrupted continuous waves.

ideal *adj.*   Assumed to fit mathematical and conceptual models that, while convenient, do not quite correspond to reality.

Ideal CONSTANTAN.

ideal bunching   In a velocity modulation tube, an idealized condition in which the electrons are arranged in bunches such that all members of a bunch have a uniform velocity and pass any given point at once.

ideal gas law   A law describing the behavior of a gas composed of particles whose dimensions are negligible in comparison to the volume which they occupy, and whose interactions apart from perfectly ELASTIC COLLISIONS are likewise negligible. At temperatures sufficiently high and pressures sufficiently low any substance will satisfy these conditions. The law states that

$$pv = RT,$$

where p is the pressure of the gas, v its volume, R the **gas constant** (which depends on the amount of gas present), and T the absolute temperature. If one MOLE of any gas is present, R (now called the **ideal gas constant**) is always about 8.314×10^7 g cm²/sec² per °K. The ideal gas constant per molecule, better known as **Boltzmann's constant,** has the value 1.3085×10^{-16} g cm²/ sec² per °K.

idiochromatic *adj.*   Indicating a crystal whose photoelectric properties are essentially those of the material itself and do not depend on impurities or imperfections.

idler frequency   The frequency at which a parametric amplifier delivers its amplified output.

IF *abbr.*   Intermediate frequency.

IGFET   INSULATED-GATE FIELD-EFFECT TRANSISTOR.

ignitor *n.*   An electron used to initiate the action of an IGNITRON.

ignitron *n.*   A rectifier tube having a mercury pool as its cathode and an ignitor inserted in the mercury as an auxiliary electrode used to control the starting of a unidirectional current flow.

ihp, i.h.p., i.hp., IHP, I.H.P. *abbr.*   Indicated horsepower.

im *abbr.*   INTERMODULATION.

image *n.*   The counterpart of an object produced by reflection, refraction, or the passage of rays through a small aperture.

image antenna   A VIRTUAL IMAGE of a real antenna that

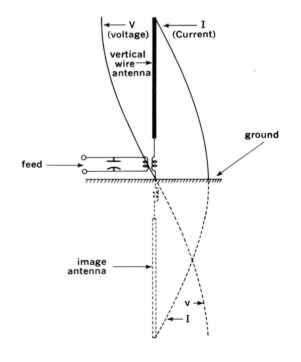

is located close to the ground, arising as a result of radiation reflected from the ground.

image frequency In a heterodyne frequency converter, an undesired input frequency capable of interacting with the local oscillator frequency in such a way as to produce a sum or difference frequency that is the same as the intermediate frequency. With respect to the desired signal frequency, image frequencies are located symmetrically about the local oscillator or intermediate frequency, whichever is higher.

image impedance The impedance of a load which will receive the maximum power when connected across a pair of terminals at which the output impedance is

$$Z = R + jX.$$

It can be shown that this impedance is

$$Z^* = R - jX,$$

that is, the CONJUGATE of Z. If both of these impedances are plotted as PHASORS on the COMPLEX PLANE, it can be seen that Z^* is the image of Z reflected across the real axis.

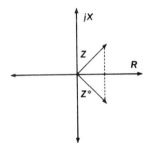

image orthicon A television camera tube in which light from the objective is directed at a photoemissive surface on the inside of the tube face. The photons emitted are accelerated towards a storage target which is then scanned by an electron beam.

image-parameter filter A FILTER consisting of a number of two-port sections connected in cascade with their IMAGE IMPEDANCES matched at their junctions. The discriminative action of the filter comes from the fact that its image impedance is a function of frequency while its terminations are resistive; the filter is thus mismatched at some frequencies.

imaginary number A number whose square is negative, that is, a number a such that $a^2 < 0$, generally written $ai = aj$, where a is any real number and $i = j = \sqrt{-1}$. See COMPLEX NUMBER.

immitance *n.* IMPEDANCE OR ADMITTANCE.

impedance *n.* **1** The total opposition, in OHMS, that a circuit offers to the flow of an alternating current, consisting of the combined effects of RESISTANCE R and inductive and capacitive REACTANCE X_L, X_C. More precisely, impedance is a complex number **Z** representing the ratio of voltage **E** to current **I**. Thus

$$Z = E/I = R + j(X_L - X_C).$$

2 By extension, in a physical system analogous to an electric circuit, the ratio of an analog of voltage to an analog of current. See ACOUSTIC IMPEDANCE.

impedance angle PHASE ANGLE.

impedance bridge A BRIDGE circuit used to measure electrical IMPEDANCE.

impedance matching The adjustment of a source impedance and an associated load impedance so that power transfer from one to the other is maximized and reflection is minimized. See CHARACTERISTIC IMPEDANCE; IMAGE IMPEDANCE.

imperfection *n.* In a crystal, any deviation from complete homogeneity of the lattice.

impulse *n.* A pulse of infinitesimal duration, that is, the sum of two step functions of opposite polarity that are separated in time by a negligible amount.

inconel *n.* An alloy typically containing 77% nickel, 15% chromium and 7% iron. Many inconels contain copper, titanium, aluminum, manganese, silicon and carbon. All have high heat-resistive strength.

increductor or **Increductor** *n.* An inductor whose value depends on the current flowing in its bias and control windings; essentially a saturable reactor designed for high frequency operation: a trade name.

incremental parameter A parameter defined as a ratio of the changes in a pair of variables rather than as a ratio of their values. For example, using Ohm's law, resistance might be defined by

$$R = \frac{E}{I};$$

incremental resistance R_i would be given by

$$R_i = \frac{dE}{dl}.$$

Such parameters are essential in analyzing the dynamic behavior of nonlinear elements, principally vacuum tubes and transistors.

index of refraction See SNELL'S LAW.

indium (In) Element 49; Atomic wt. 114.76; Melting pt. 155°C; Boiling pt. 2000 = ± 10°C; Spec. grav. (20°C) 7.28; Hardness 1.2; Valence 1 or 3; Spec. heat 0.057 cal/(g)(°C); Electrical conductivity (0–20°C) 0.111 × 10⁶ mho/cm. Used in sleeve bearings for engines, in low-melting alloys for fusible safety links, and as an acceptor impurity in semiconductors. Its phosphate is used in high-temperature transistors and solar batteries, and its arsenide and antimonide are used in low-temperature transistors.

Indox *n.* The trade name of a series of barium FERRITES with extremely high coercive force.

induce *v.* To produce by induction, as an electric current or magnetic effect.

inductance *n.* **1** The property of an electric circuit (or of two circuits in close proximity) by which, through electromagnetic INDUCTION, a change in the current in one circuit produces an ELECTROMOTIVE FORCE, either in itself or in the neighboring circuit. See MUTUAL INDUCTANCE; SELF INDUCTANCE. **2** An ideal inductor.

induction *n.* **1** The production of an electric charge or magnetic field in a body as a result of the application of an electric or magnetic field. See ELECTRIC INDUCTION; MAGNETIC INDUCTION. **2** The production of an electromotive force in a conductor either as a result of its motion through a magnetic field in such a way as to cut across magnetic flux or as a result of a change in the magnetic flux surrounding it. In the first case the electromotive force E is given by

$$E = \int_c (\mathbf{B} \times \mathbf{v}) \cdot d\mathbf{l},$$

where **B** is magnetic induction, **v** velocity, *dl* an infinitesimal length of the conductor, and the indicated integral is taken along the conductor. The second case is described by FARADAY'S LAW OF ELECTROMAGNETIC INDUCTION.

induction coil A device consisting of two concentric coils and an INTERRUPTER, that changes a low steady voltage into a high intermittent alternating voltage by electromagnetic induction.

induction heating The heating of a conductive material by placing it in an electromagnetic field that varies with time, heat being produced by the flow of induced currents through the internal resistance of the material.

induction motor A type of ac motor in which the current in the secondary winding, most often the rotor, is induced by the changing magnetic field created by the current in the primary winding.

inductive *adj.* Of or having to do with inductance, induction, or an inductor.

inductive reactance See REACTANCE.

inductometer *n.* An instrument for measuring inductance.

inductor *n.* A device designed to introduce INDUCTANCE into a circuit or network.

inertance *n.* The acoustic or mechanical analog of inductance. See ACOUSTIC IMPEDANCE.

information *n.* That property of a signal or message whereby it conveys something unpredictable by and meaningful to the recipient, usually measured in bits.

information theory The mathematical study and analysis of the problems of encoding and decoding messages, based heavily on the science of probability. Through its ability to measure information quantitatively, this study provides criteria for evaluating the performance of communications systems.

infra- *prefix* Below; beneath; less than.

infrared *adj.* Having wavelengths greater than that of a visible red light and shorter than microwaves.

infrared radiation Electromagnetic radiation having wavelengths between about 0.75 micron and 1000 microns, that is, between those of microwaves and visible red light. This range is sometimes divided into **near infrared** (0.75–3.0 microns), **middle infrared** (3.0–30.0 microns), and **far infrared** (30–1000 microns).

infrasonic *adj.* Having a frequency below 20–25 hz.

inhibitor or **inhibition gate** In a digital computer, a LOGIC CIRCUIT that clamps a specified output to the zero level when energized.

injection grid In the mixer tube of a heterodyne frequency converter, a grid through which the local oscillator signal is introduced.

input *n.* **1** The signal applied to a network or device. **2** The energy supplied to a system or device. **3** The terminals or PORT through which an input is supplied. **4** The information or data fed to a computer for processing.

input admittance See DRIVING-POINT IMMITANCE.

input gap The pair of electrodes that modulates the electron beam in a velocity-modulation electron tube.

input impedance See DRIVING-POINT IMMITANCE.

insertion gain (loss) The GAIN (LOSS) that results from the insertion of a transducer into a system, generally expressed in decibels as the ratio of the power available at the point in the system fed by the transducer after its insertion to the power available at that point before its insertion. If at any point the signal consists of more than one component the components should be specified.

insertion-loss filter A filter specified and designed on the basis of its insertion loss for various frequency components. In a low-pass filter, for example, insertion loss in the pass band, $0 < \omega < \omega_1$ would be required to remain below a certain maximum, while the insertion loss in the stop band, $\omega_2 < \omega < \infty$, would be required to remain above a certain minimum.

instruction *n.* In a digital computer, a set of characters that define an operation and the ADDRESSES of the OPERANDS upon which the operation is to be performed.

insulate *v.* To surround or separate with nonconducting material in order to prevent or lessen the leakage of electricity, heat, sound, etc.

insulated-gate field-effect transistor A METAL-OXIDE SEMICONDUCTOR FIELD-EFFECT TRANSISTOR.

insulator *n.* **1** A material having a very low electrical conductivity and high dielectric strength. **2** A structure made of such a material, used to support an energized conductor while keeping it electrically isolated.

integral 1 *n.* An operator, which operates on the continuous function $f(x)$ and is written symbolically

$$\int f(x)dx.$$

Geometrically an integral can be interpreted as the

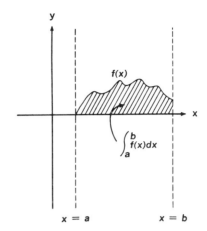

area between the curve $f(x)$ and the x axis (see illustration). In the case where the integration takes place over a finite interval $a \le x \le b$ the integral is termed a **definite integral** and

$$\int_a^b f(x)dx$$

equals a numerical quantity. More precisely, if the interval $[a,b]$ is divided into n subintervals Δx_i (not necessarily all equal), and an x_i is chosen in each Δx_i and a sum

$$S = \sum_{i=1}^{n} f(x_i)\,\Delta x_i$$

is formed, the integral

$$\int_a^b f(x)dx$$

is the limit of S as n approaches infinity in such a way that the largest x_i approaches zero, provided that this limit exists. It can be shown that if $F(x)$ is the PRIMITIVE of $f(x)$, then

$$\int_a^b f(x)dx = F(b) - F(a).$$

If the integral is unbounded

$$\int f(x)dx$$

is termed an **indefinite integral** and the result is a function of x. Integrals occur commonly in electronics; for example, the voltage V across a capacitor C is given by

$$v = 1/C \int_{t_1}^{t_2} i\,dt,$$

where i is the current and t time. **2** *adj.* Of, in the form of, or having to do with an integer.

integrate *v.* To compute the INTEGRAL of (a FUNCTION). —**integration** *n.*

integrated circuit A small chip of solid material (generally a semiconductor) upon which, by various techniques, an array of active and/or passive components have been fabricated and interconnected to form a functioning circuit. Integrated circuits, which are generally encapsulated with only input, output, power supply, and control terminals accessible, offer great advantages in terms of small size, economy, and reliability.

integrator *n.* A device or network whose output is the INTEGRAL with respect to time of its input, or a close approximation thereof.

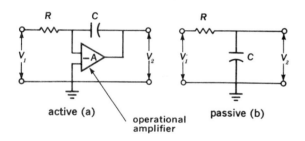

active (a)   passive (b)

operational amplifier

(a) If |A| ≫ 1

$$V_2 \cong -\frac{1}{RC} \int_0^t V_i \, d_\tau$$

(b) If RC ≫ 1

$$V_2 \cong \frac{1}{RC} \int_0^t V_i \, d_\tau$$

intellectronics *n.* A branch of electronics concerned with the development of data-processing machines that are able to alter their own programming in such a way as to become increasingly efficient and adaptable, thus displaying qualities comparable to human intelligence and learning ability.—**intellectronic** *adj.*

intelligence *n.* INFORMATION.

intensifier *n.* In a CATHODE-RAY TUBE, an electrode that further accelerates the electron beam after it has been deflected.

intensity *n.* **1** The strength of an electric or magnetic field. **2** Current strength, in amperes. **3** Electromotive force; potential.

interbase resistance In a UNIJUNCTION TRANSISTOR, the resistance between base$_1$ and base$_2$.

intercarrier sound In television receivers, a technique in which the sound and video carriers are heterodyned, producing a frequency-modulated difference signal that contains the sound information.

interface resistance In an electron tube, a parasitic resistance that develops in the cathode circuit as a result of the formation of an interface layer between the base metal of the cathode and the emissive surface.

interferometer *n.* Any of various instruments in which radiation from a single source is split into two or more beams which are recombined after traversing paths of differing lengths or passing through different media. Observation and measurement of the resulting interference patterns permit very accurate determinations of wavelength, distance, refractive index, etc.

interlaced scanning A system, used in television, in which 30 frames (pictures) are presented each second but each frame is presented in two fields, one in which the odd-numbered lines are scanned and another in which the even-numbered lines are scanned. This is done to suppress visible flickering.

intermediate frequency In a superheterodyne radio receiver, the DIFFERENCE FREQUENCY resulting from the HETERODYNE action of the incoming signal frequency and the local oscillator frequency, normally the frequency at which most of the signal amplification takes place.

intermodulation *n.* The modulation of the components of a complex wave one by the others as a result of HETERODYNE action in some nonlinear element of the signal path.

international ampere The current which on passing through a silver-nitrate solution will deposit silver at the rate of 0.001118 gram per second.

international coulomb The quantity of electric charge whose passage through a surface in 1 second constitutes a current of 1 international ampere.

international ohm The resistance at 0°C of a uniform column of mercury having a mass of 14.4521 grams and a length of 106.3 centimeters.

international volt The potential difference which, when applied to a conductor having a resistance of 1 international ohm, produces a current of 1 international ampere.

interphase transformer In certain polyphase rectifier systems, a transformer, often an autotransformer with a center tap, used to connect points whose average dc potential is the same but whose ac potentials are different.

interpole *n.* An auxiliary pole placed between the main poles of a rotary dc machine to provide additional flux to assist commutation.

interrogate *v.* To transmit a signal to (a remote component of a system) that causes it to respond in some way, as by transmitting stored data, an identification signal, etc.—**interrogation** *n.*

intrinsic semiconductor A semiconductor material in which the density of charge carriers (ELECTRONS and HOLES) depends not on the concentration of impurities and imperfections but on the material itself. The available charge carriers are thermally generated electrons and holes in approximately equal numbers.

intrinsic standoff ratio In a UNIJUNCTION TRANSISTOR, a parameter η given by $\eta = R_{B1}/R_{BB}$, where R_{B1} is the resistance of base$_1$ and R_{BB} is the INTERBASE RESISTANCE.

Invar *n.* An alloy of iron and 34-36% nickel with a trace of carbon having an extremely low coefficient of expansion ($.9 \times 10^{-6}$) at 0°C at atmospheric temperatures, used in bimetals, thermostats, and length standards.

inverse feedback Negative FEEDBACK.

inverter *n.* **1** A device that converts dc to ac. The term *inverter* generally implies that this conversion is accomplished by strictly electrical and not mechanical means. **2** A LOGIC CIRCUIT that produces an output when deenergized and no output when energized. See NAND-GATE; NOR-GATE; NOT-GATE.

iodine (I) Element 53; Atomic wt. 126.91; Melting pt. 113.5°C; Boiling pt. 184.35°C; Spec. grav. (20°C solid) 4.93; Density (gas) 11.27 g/l; Valence 1, 3, 5, or 7; Heat of fusion 3770 cal/mole; Heat of vaporization 9970 cal/mole; Spec. heat 0.052 cal/(g)(°C); Electrical conductivity (0–20°C) 10^{-6} mho/cm.

ion *n.* The charged particle formed when one or more electrons are taken from or added to a previously neutral atom or molecule.

ionize *v.* To change or dissociate into IONS. —**ionization** *n.*

ionizing event An event leading to the formation of ions.

ionosphere *n.* The region of the earth's atmosphere from about 30 to 250 miles up, consisting of several ionized layers whose heights and intensity of ionization vary seasonally and diurnally. It is usually divided into three regions: D-REGION; E-REGION; F-REGION.

ionospheric wave A radio wave that propagates by successive reflections from the ionosphere.

ion trap A device used in a cathode-ray tube to prevent ions from striking and damaging the phosphor coating.

iota (written I, ι) The ninth letter of the Greek alphabet, used symbolically in lower case to represent a unit vector.

ips *abbr.* Inches per second.

iridium (Ir) Element 77; Atomic wt. 193.1; Melting pt. 2350°C; Boiling pt. above 4800°C; Spec. grav. (17°C) 22.42; Hardness 6.0-6.5; Valence 3 or 4; Spec. heat 0.031 cal/(g)(°C); Electrical resistivity (0°C) 5.3 $\times$ 10^{-6} ohm/cm. Used to strengthen platinum alloys, for very high temperature crucibles operating in non-oxidizing atmospheres, and in furnace windings.

iron (Fe; *Lat.* **ferrum.)** Element 26; Atomic wt. 55.85; Melting pt. 1535°C; Boiling pt. 3000°C; Spec. grav. (20°C) 7.85-7.88; Hardness (iron) 4-5, (steel) 5-8.5; Valence 2, 3, or 6; Spec. heat 0.11 cal/(g)(°C); Electrical conductivity (0–20°C) 10^5 mho/cm. Used in ferromagnetic alloys, both for permanent magnets and for cores of transformers, inductors, etc. Iron is an important constituent of FERRITES, and is also used in several high-resistivity alloys.

iron-silicon alloy A soft magnetic alloy used in transformers to cut hysteresis loss.

IS *abbr.* Internal shield.

iso- *combining form* Equal.

isochrone *n.* A line or surface whose points are all equal in some sense with respect to time, in particular, a line or surface along which the radiation from a source arrives with an identical time delay at all points.

Isoperm *n.* A soft magnetic alloy of 60% iron and 40% nickel having high and constant permeability.

isostatic *adj.* Having to do with or characterized by equilibrium resulting from equal pressure on all sides.

isotope *n.* Any of a set of nuclides having the same atomic number (and therefore the same identity as an element) but different mass numbers. As the stability of a nucleus depends on its ratio of mass to charge, some isotopes are radioactive.

isotropic *adj.* Exhibiting the same physical properties in all directions.

iterative impedance In a two-port network, the impedance that when connected across one pair of terminals causes the same impedance to appear between the other pair of terminals.

ITU *abbr.* International Telecommunication Union.

I.V., i.v. *abbr.* Initial velocity.

J

J *symbol* **1** (*J*) JOULE. **2** (**J**) Electric current density. **3** (*j*) Square root of −1. **4** (**j**) Unit vector along the y-axis. **5** (*J*) Emissive power.

jack *n.* A connecting device provided with metal spring clips to which the wires of a circuit can be connected and into which a fitting plug can be inserted.

jag *n.* In facsimile transmission, distortion that results from a temporary loss of synchronization between scanner and recorder.

jam *v.* To interfere electronically with the reception of radio signals.

J-display In radar, an A-DISPLAY modified in such a way that the time base appears as a circle, the amplitude of the target pulses being measured radially outward.

jitter *n.* Instability of a waveform, resulting from fluctuations in supply voltages, components, etc.

joule *n.* In the MKS system, the measure of work or

energy, defined as equal to the work done by a force of 1 NEWTON in producing a displacement of 1 meter in the direction of the force. One joule equals one watt-second.

Joule's law A law stating that the amount of heat *H* produced in a conductor by the flow of a steady current *I* is given by

$$H = KI^2Rt,$$

where *R* is the resistance of the conductor, *t* is the time for which the current flows, and *K*, the constant of proportionality, has the value 0.2390 calories per joule when *R* is in ohms and *I* in amperes.

jumper *n.* A conductor used to make a temporary bypass or other connection.

junction *n.* **1** In an electrical network, a NODE that is common to three or more elements. **2** In a waveguide, a point at which energy may flow into two or more paths. **3** In a solid-state device, a region in which there is a transition between semiconductor regions whose electrical properties differ. **4** A point at which dissimilar materials, as the metals of a thermocouple, make contact.

junction barrier A DEPLETION REGION.

junction diode A semiconductor diode whose characteristics result largely from the fact that it contains a PN JUNCTION.

junction field-effect transistor A FIELD-EFFECT TRANSISTOR in which the gate and channel form a PN JUNCTION.

junction transistor See TRANSISTOR.

K

K *symbol* **1** (**k**) capacity. **2** (K) Cathode. **3** (K) Kelvin. **4** (k-) kilo-. **5** (K) Kilohm. **6** (K) Potassium. **7** (**K**) Relay.

kA *abbr.* Kiloampere(s).

kanthal A An alloy of 69% iron, 23% chromium, 2% cobalt, 6% aluminum. High resistivity alloy; oxidation-resistant to about 1315°C.

kappa (*written* K, κ) The tenth letter of the Greek

alphabet, used symbolically to represent: **1** (κ) Curvature. **2** (κ) Susceptibility. **3** (κ) Coupling coefficient.

kb *abbr.* Kilobar(s).

K band A band of microwave frequencies extending from 11 ghz (11×10^9hertz) to 36 ghz.

kc *abbr.* **1** Kilocycle. **2** Kilocycles. **3** Kilocycles per second.

kc/s, kc/sec *abbr.* Kilocycles per second.

K display In radar, a modification of an A DISPLAY in which each waveshape is duplicated, differences in height of the two waveshapes indicating the amount of error in aiming.

Kelvin bridge An adaptation of the WHEATSTONE BRIDGE, used to measure very low resistances, offsetting the parasitic resistance of leads and contacts. In the balanced bridge illustrated the unknown resistance (R_x) will be given by

$$R_x = \frac{R_B}{R_A} R_s + R_y \left(\frac{R_b}{R_a + R_b + R_y} \right) \left(\frac{R_B}{R_A} - \frac{R_b}{R_a} \right)$$

As R_s is in series with R_a, any error in R_s can be considered dR_a. Thus, by making $R_B/R_A - R_b/R_a$ as close to zero as possible, making R_a as large as is convenient, and making R_y as small as possible the second term of the equation (which contains the error) can be made negligible.

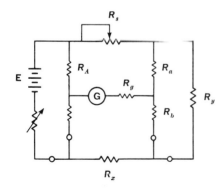

Kelvin scale A modification of the Celsius (centigrade) scale of temperature measurement in which absolute zero is taken as zero. 0°C is thus 273.15°K.

Kennelly-Heaviside layer HEAVISIDE LAYER.

kenopliotron *n.* A vacuum diode and triode combined in one envelope in such a way that the same element acts as the anode of the diode and the cathode of the triode.

kenotron *n.* A thermionic vacuum diode, used principally as a high-voltage rectifier.

kernel *n.* The function $K(x, t)$ in an integral equation of the form

$$f(x) = \int_a^b K(x, t)y(t)dt.$$

Kerr cell A device in which the KERR ELECTRO-OPTIC EFFECT may be observed, consisting of a glass cell containing a suitable dielectric fluid which can be subjected to an electric field by means of a pair of capacitor plates included in the cell.

Kerr-electro-optic effect The anisotropy and birefringence exhibited by certain materials when they are subjected to an electric field. The effect arises from a partial orientation that the electric field imposes on the molecules of the material.

key *n.* A hand-operated switch used to control an electrical signal, as in the transmission of code.—**key** *v.* —**keyer** *n.*

key click A TRANSIENT resulting from the opening or closing of a key.

keystone distortion A form of distortion in which the rectangular picture pattern in a television tube becomes trapezoidal.

kg *abbr.* Kilogram(s).

kG *abbr.* **1** Kilogauss. **2** Kilogausses.

kgf *abbr.* Kilogram-force.

kg-m *abbr.* Kilogram-meter.

khz *abbr.* Kilohertz.

kickback *n.* The electromotive force that appears across an inductance when the current through it is suddenly reduced. See COUNTERVOLTAGE.

kilo- *prefix* One thousand times (a specified unit).

kinerecording *n.* The method of preserving a television program by photographing a monitor set with a cinecamera.

kinescope *n.* A cathode ray tube used as the video display of a television receiver.

kinetic energy The energy E that a body has by virtue of the fact that it is in motion, given by

$$E = \tfrac{1}{2}(mv^2 + I\omega^2),$$

where m is the mass of the body, v its linear velocity, I its moment of inertia, and ω its angular velocity. For very large velocities correction according to the theory of relativity is necessary.

Kirchhoff's laws of electrical networks Either of a pair of laws that state certain general restrictions on the current and voltage in any electrical NETWORK. These laws may be stated as follows: **1** At each instant of time the sum of the voltages around any LOOP is zero. **2** At any particular instant the sum of the currents flowing to a point equals the sum of the currents flowing away from that point.

Kirchhoff's radiation law A law stating that at any given temperature and for the same kind of radiation the absorptivity and emissivity of a body are equal.

kJ, kj *abbr.* Kilojoule(s).

klystron *n.* An electron tube in which a velocity-modulated electron beam is used to amplify radio-frequency signals, energy being coupled into and out of the beam by means of resonant cavities. The klystron is useful from about 200 Mhz to 30 ghz.

K Monel An age-hardenable alloy of 66% nickel, 29% copper, 2.75% aluminum, 0.9% iron, 0.75% manganese, 0.15% carbon, 0.5% silicon, and 0.005% sulfur, having high corrosion resistance.

knee *n.* A section between two comparatively straight segments of a curve in which the magnitude of curvature, although of the same sign, is relatively high.

knife switch A switch consisting of a knifelike blade which can be pushed between spring contacts.

knockout *n.* A portion of a metal cabinet or outlet which can be easily removed for the attachment of cables or fittings.

Konal metal An alloy of nickel, cobalt, iron and titanium used as a sleeve in the cathodes of radio receiving tubes.

Kovar *n.* An alloy of iron, nickel (23–30%), cobalt (15–23%), and manganese (0.6–0.8%) having the same expansion properties as glass and used for glass to metal seals in electronic tubes. Specific gravity 8.2; thermal conductance 0.193 watts/cm°C; melting point 1450°C. A trademark.

krypton (Kr) Element 36; Atomic wt. 83.8; Melting pt. −157°C; Boiling pt. −152.9°C; Density (0°C) 3.708 g/l; Valence 3. Used as an atmosphere in electric lamps, electric-arc lamps, and other electronic devices.

kV, kv *abbr.* Kilovolt(s).

kVA, kva *abbr.* Kilovolt-ampere.

kVAH, kvah *abbr.* Kilovolt-ampere hour.

kVAr, kvar *abbr.* Kilovar.

kw *abbr.* Kilowatt(s).

kWh, kwhr, K.W.H. *abbr.* Kilowatt-hour.

L

l *abbr.* Length.

L *symbol* **1** (L) Inductance. **2** (L) Inductor. **3** (l) Lambert.

LA *abbr.* Level amplifier.

labile oscillator An oscillator whose frequency is remote-controlled.

ladder network A network consisting of a set of H, L, T, or pi networks connected in cascade.

lag *n.* **1** The amount by which the phases of an alternating quantity follow the corresponding quantity of the same frequency, that is, if

$$f(t) = \cos(\omega t)$$

and

$$g(t) = \cos(\omega t + \phi),$$

then $g(t)$ will have a lag of ϕ compared with $f(t)$. **2** The persistence of the electric charge image in a television camera tube.

lag circuit An LC (inductance-capacitance) low-pass filter placed between a key and a keyed circuit in order to suppress key clicks and delay the response of the keyed circuit.

lambda (*written* Λ, λ)   The eleventh letter of the Greek alphabet, used symbolically to represent any of various coefficients, constants, etc., of which in electronics the principal one is (λ) wavelength in meters.

lambert *n.*   A measure of brightness, derived from the brightness of a perfectly diffusing surface emitting or reflecting light at the rate of one lumen per cm².

land *n.*   **1** On a printed circuit board, the conductive areas to which connections are made and components attached. **2** The surface between the grooves on a recording disk.

land line   An electrical TRANSMISSION LINE which passes over land.

Langmuir dark space   The nonluminous area in the plasma of a glow discharge tube created when a negative probe is inserted.

lanthanum (La)   Element 57; Atomic wt. 138.92; Melting pt. 826°C; Boiling pt. 1800°C; Spec. grav. 6.155; Valence 3; Spec. heat 0.045 cal/(g)(°C); Electrical conductivity (0–20°C) 0.017 × 10⁶ mho/cm.

lap *n.*   A rotating plate covered with liquid abrasive used for grinding quartz crystals.—**lap** *v.*

Laplace transform   A linear operator that can be used to transform a linear differential equation with constant coefficients and its initial conditions into an algebraic equation whose roots can be inversely transformed into particular solutions of the original differential equation. Analytically, the single-sided Laplace transform (so called because of the choice of the limits of integration) is given by

$$\mathcal{L}(F) = f(s) = \int_0^\infty F(t)e^{-st} \, dt,$$

and the inverse transform

$$\mathcal{L}^{-1}(f) = F(t) = \int_{c-j\infty}^{c+j\infty} f(s)e^{st} \, dt.$$

The functions $F(t)$, $f(s)$ are called **Laplace transform pair.** The linear properties of the Laplace transform makes it convenient to present transforms of various functions in tables. There are restrictions as to the functions to which the technique may be applied.

large signal   A signal sufficiently large that the EQUIVALENT CIRCUIT of an inherently nonlinear device, (as a transistor or vacuum tube) to which it is applied is grossly inaccurate unless nonlinear elements are included. Circuit analysis under these conditions is far more complicated than for SMALL SIGNALS.

Larmor frequency   The angular frequency ω_L with which a charged particle in rotation precesses about a steady magnetic field, given by

$$\omega_L = (q/2m)B,$$

where q is the charge of the particle, m its mass, and B the magnetic field strength.

laser *n.*   A device in which the excitation energy of resonant atomic or molecular systems is used to coherently amplify or generate light. Consider a system of particles with energy levels E_1, E_2; where $E_1 < E_2$. A particle passing between these levels absorbs or emits a quantum of radiation of frequency

$$\nu = (E_2 - E_1)/h,$$

where h is Planck's constant. If N_1 is the number of particles in E_1 and N_2 the number in E_2, it follows from the laws of thermodynamics that at a positive absolute temperature $N_2 < N_1$. As a result of this, a wave of the appropriate frequency would be attenuated. If by some means N_2 can be made to exceed N_1 by a reasonably large margin, an incoming wave of frequency ν will stimulate transitions from E_2 to E_1 and in the process become coherently amplified. For sustained oscillation to occur feedback must be provided; this is often done by covering the ends of the chamber in which the reaction occurs with reflective material and arranging the chamber to be resonant at the frequency of oscillation.

lattice *n.*   The geometrical arrangement of atoms in a crystal.

lattice network   A network of four branches connected in a loop, with input terminals on two non-adjacent junction points and output terminals on the other two.

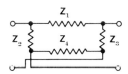

lattice winding   HONEYCOMB WINDING.

lavalier microphone   A small microphone hung from the user's neck.

lb., lbs. *abbr.* Pound(s).

L band A band of microwave frequencies extending from 390 Mhz to 1,550 Mhz.

L display In radar, a modification of an A display in which the target appears as two waveshapes on opposite sides of the central time base, differences in amplitude of the waveshapes indicating the amount of error in aiming.

lead (Pb; *Lat.* plumbum**)** Element 82; Atomic wt. 207.21; Melting pt. 327.4°C; Boiling pt. 1620°C; Spec. grav. (20°C) 11.35; Hardness 1.5; Valence 2 or 4; Spec. heat 0.031 cal/(g)(°C); Electrical conductivity (0–20°C) 0.046 × 10⁶ mho/cm. Used in various alloys, as solders, type metal, pewter, etc.; also used for storage battery plates, and as a shielding against radioactivity.

lead *n.* **1** A wire connecting two circuit elements. **2** A negative LAG.

lead-in *n.* **1** The blank spiral groove at the start of a recording disk. **2** A DOWN-LEAD.

lead screw A threaded shaft guiding the cutter of a disk recorder to space the grooves accurately.

leakage current An unwanted flow of current.

leakage flux The magnetic lines of force that extend beyond the area in which they are useful.

leakage inductance The inductance caused by LEAKAGE FLUX inside the windings of a transformer.

leakage reactance In a transformer, the part of the inductive reactance of a coil that results from magnetic flux that does not link with another coil.

leakage resistance The resistance, ideally infinite, of the path taken by the LEAKAGE current through a device or component.

leaky *adj.* Having leakage.

Lecher oscillator A device in which LECHER WIRES are used to measure wavelength.

Lecher wires A type of transmission line used to measure wavelength, consisting of a pair of wires whose electrical length is adjustable. If a radio-frequency source is coupled to one end of the line and the line is adjusted until a set of standing waves are formed, the wavelength may be determined by measurement of the distance between adjacent nodes.

Leclanche cell A carbon-zinc dry cell.

left hand rule A modification of FLEMING'S RULE, in which the left hand is employed instead of the right. The difference is that the thumb (def. 1) or index finger (def. 2) is pointed in the direction of electron flow instead of current flow.

Lenard rays Cathode rays that emerge into the air out of a vacuum tube constructed with especially thin glass or metal at the emergence point.

lens *n.* A device designed to make some form of radiation converge or diverge as a result of being refracted. The most common and familiar lenses are those designed to operate on light, but analogous devices exist which operate on radio waves, sound waves, and particulate radiation such as electron beams.

Lenz's law A law which states that if a current is induced in a circuit due to its motion in a magnetic field or to a change in its magnetic flux, the direction of the induced current will be such as to exert a magnetic force opposing the motion or change in flux.

LET *abbr.* Linear energy transfer.

Leyden jar The earliest form of capacitor, consisting of a glass jar lined on the inside and outside with metal foil.

Lf *abbr.* Low frequency.

lifetime *n.* **1** MEDIAN LIFE. **2** In a semiconductor, the amount of time a hole or electron exists before undergoing recombination.

light *n.* Energy in the form of electromagnetic radiation that is capable of creating visual responses in the human eye, being generally between about 4000 angstroms (violet) and 7700 angstroms (red) in wavelength. Sometimes the term is understood to include INFRARED RADIATION, ultraviolet radiation, and other radiation as well.

light-negative Decreasing in electrical conductivity when exposed to light.

lightning arrester A device to prevent damage to electrical equipment by transient overvoltages whether from lightning or switching. Spark gaps which can only be bridged by voltages above those used in the equipment allow the higher voltages to be discharged to ground.

light pipe A bundle of transparent fibers which can

transmit light around corners with small losses. Each fiber transmits a portion of the image through its length, reflection being caused by the lower refractive index of the surrounding material, usually air. Each fiber is of the order of 0.0005 in. in cross section; thus a 525-line television image can be transmitted in a ¼ in. pipe.

light-positive   Increasing in electrical conductivity when exposed to light.

light valve   A device whose transmission of light can be varied by electrical or electromagnetic means.

limen *n.*   The minimum excitation necessary to produce a particular response.

limit *n.*   In mathematics, **1:** A finite number related to a sequence S_n in such a way that there exists a number N for which $|L - S_n| < \varepsilon$, when $n > N$, ε being an arbitrarily small positive number. Symbolically,

$$L = \lim_{x \to \infty} S_n.$$

2 A finite number L related to a function $f(x)$ in such a way that there exists a number δ for which $|L - f(x)| < \varepsilon$, whenever $|x - a| < \delta$, ε being an arbitrarily small positive number. Symbolically,

$$L = \lim_{x \to a} f(x).$$

limiter *n.*   A circuit that prevents the amplitude of a waveform from exceeding a certain level, preventing swings in either a positive or negative direction, or both.

line *n.*   **1** A TRANSMISSION LINE. **2** The path of an electron beam in one horizontal sweep across a television picture tube. **3** A horizontal scanning element in a facsimile system. **4** TRACE. **5** MAXWELL.

line amplifier   An amplifier whose function it is to deliver an audio signal to a transmission line, generally at a predetermined level.

linear *adj.*   Having behavior or a response that can be completely described by a set of LINEAR EQUATIONS or a set of LINEAR DIFFERENTIAL EQUATIONS. —**linearity** *n.*

linear differential equation   A DIFFERENTIAL EQUATION that is of the first degree in the dependent variable and any of its derivatives that are present, that is, an equation of the form

$$\sum_{i=0}^{n} p_i(x) f^{(i)}(x) = Q(x),$$

where p_i is a polynomial in x.

linear equation   An algebraic equation in which all of the variables are present in the first degree only.

linear programming   A part of mathematics that is concerned with finding the conditions for which a sum of the form

$$S = \sum_{n=1}^{m} c_i x_i$$

is maximized (or minimized), where c_i are a set of constants, and x_i a set of variables upon which certain constraints have been imposed.

line loss   The energy lost in a transmission line.

line microphone   A unidirectional microphone consisting of several tubes of different lengths pointing toward the source of sound and carrying soundwaves to the transducer.

line radiator   A number of point sources of radiation closely spaced along a line, often approximated for acoustic purposes by a number of identical radiators arranged in a line, all operating in phase.

link *n.*   A radio system sending signals between two stations.

linkage *n.*   FLUX LINKAGE.

link fuse   An open strip of fusible material attached to terminal blocks.

liquid rheostat   A rheostat in which the resistance is controlled by the level of a conductive liquid between metal plates.

Lissajous figure   A pattern resulting from the super-

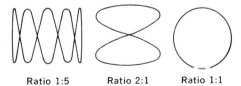

Ratio 1:5   Ratio 2:1   Ratio 1:1

Typical Lissajous figures for sine waves of the frequency ratios indicated.

position of two harmonic motions at right angles, as when two sine waves whose frequencies are integrally related are applied to the horizontal and vertical deflection circuits of a cathode ray tube.

lithium (Li)　Element 3; Atomic wt. 6.940; Melting pt. 186°C; Boiling pt. above 1220°C; Spec. grav. (20°C) 0.534; Valence 1; Heat of fusion (179°C) 103.2 cal/g; Heat of vaporization (1317°C) 4680 cal/g; Viscosity (200°C) 5.62 millipoises, (400°C) 4.02 millipoises, (600°C) 3.17 millipoises; Vapor pressure (727°C) 0.78 mm; Thermal conductivity (216°C) 0.109 cal/(sec)(cm²)(cm)(°C); Surface tension (200–500°C) About 400 dynes/cm; Heat capacity (100°C) 0.90 cal/(g)(°C); Electrical resistivity (0°C) 1.34×10^{-6} ohm/cm, (100°C) 12.7×10^{-6} ohm/cm. Used on cathodes of phototubes; its hydroxide is used as an additive in alkaline storage batteries.

litz wire　A wire formed of very fine, separately insulated, conducting strands, so interwoven that each passes back and forth between the center and the outside, thus reducing SKIN EFFECT.

live *adj.*　Connected to a source of electrical energy; energized.

LLL *abbr.*　Low level logic.

lm/FT² *abbr.*　Lumens per square foot.

lm-hr *abbr.*　Lumen-hour.

lm/m² *abbr.*　Lumens per square meter.

lm/W *abbr.*　Lumen per watt.

ln *abbr.*　Natural logarithm.

L network　A two-port network of two branches connected in series with the free ends connected to one pair of terminals and the junction and other free end connected to another pair.

load *n.*　**1** The amount of electrical power drawn from a particular source. **2** A device that absorbs electric power. **3** The impedance to which energy is being applied.

load *v.*　**1** To draw power from. **2** To add certain qualities (for example, impedance, inductance or admittance) to a system in order to give it certain desired characteristics. **3** In a computer, to transfer information into the internal storage.

load curve　**1** A plot against time of the power drawn by

a LOAD. **2** The ellipse formed by a LOAD LINE when the load is reactive.

load diagram　For an electronic oscillator, a plot of the frequency and power output as a function of the load IMMITTANCE.

loading coil　**1** A PUPIN COIL. **2** An inductance used to increase the electric length of an antenna and thus lower its resonant frequency.

load line　A line drawn across a set of characteristic curves of a nonlinear device, as a transistor or vacuum tube, in such a way as to relate a pair of operating parameters. In this way the values of these parameters at various operating points can be determined.

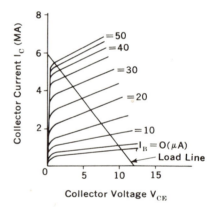

Load line for a transistor

lobe *n.*　In a RADIATION PATTERN, one of the areas of greater radiation. The axis of the lobe indicates the direction of greatest response. The largest lobe is called the **major lobe** all others are **minor lobes**.

local oscillator　In a HETERODYNE FREQUENCY CONVERTER, the oscillator which supplies to the MIXER the frequency needed to heterodyne the incoming signal frequency to the new output frequency.

lock 1 *v.*　To synchronize or become synchronized with; follow or control precisely, as in frequency, phase, motion, etc.: used with *in, on, onto*, etc. **2** *n.* Something that locks.

lockover circuit   A BISTABLE circuit.

loctal *n.*   A type of electron tube base characterized by a central guidepost with an aligning groove and eight pins piercing the tube's glass seal.

lodar *n.*   In LORAN, a method of excluding the effect of reflected waves from the ionosphere by means of a loop antenna directed to give a null indication of particular components.

lodestone *n.*   MAGNETITE.

Lodex *n.*   A magnetic material consisting of small iron-cobalt magnets dispersed in a non-magnetic lead matrix, allowing easy machining.

log *n.*   A register of operations, required by law from radio operators.—**log** *v.*

logarithm *n.*   A number *m* related to another number *p* by

$$B^m = p,$$

where *B* is an arbitrarily chosen number larger than 1. Usually *m* is written $\log_B p$. Logarithms have the basic properties that

$$\log pq = \log p + \log q$$

and

$$\log p^q = q \log p.$$

In the systems most often used B = 10 (**common logarithms**) or B = *e* = 2.71828 . . . (**natural logarithms**). The common logarithm of *p* is written $\log_{10} p$ or simply log *p*, the natural logarithm of *p* is written $\log_e p$ or ln *p*. —**logarithmic** *adj.*

logger *n.*   A device for measuring and recording quantities automatically.

logic *n.*   In a digital computer, the arrangement of circuits which perform the basic operations.

logic circuit   Any of various switching circuits, as AND-gates OR-gates, etc., that can perform LOGIC OPERATIONS or represent LOGIC FUNCTIONS.

logic operation or **logic function**   In BOOLEAN ALGEBRA, any operations designated by AND, OR, NOR, etc.

log-periodic antenna   Any of a class of broadband antennas whose electrical properties are periodic with the NATURAL LOGARITHM of frequency. The geometrical shapes used in such antennas are, in general, based on or related to shapes such as the plane curve given by $\theta = k \sin (\ln r)$. This principle can also be used to determine the lengths and spacings of dipole and other arrays.

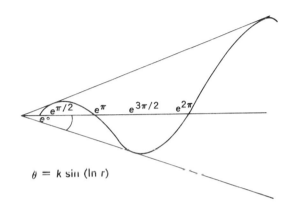

$\theta = k \sin (\ln r)$

longevity *n.*   The normal operating lifetime of a piece of equipment, usually considered to be the period in which its failure rate is acceptably low and essentially constant.

longitudinal wave   A wave in which the direction of the displacement of the medium at each point of the wavefront is normal to the wavefront.

long play record   A recording disk, 10 or 12 inches in diameter, with fine grooves and a speed of 33⅓ or 16⅔ rpm.

long wave   A radio wavelength of more than 1,000 meters with frequencies of less than 300 khz.

long-wire antenna   An antenna consisting of conductors that are long, usually by a number of wavelengths, at the operating frequency.

loop *n.*   **1** In an electrical network, a closed path through two or more circuit elements. **2** In a digital computer program, a sequence of instructions that is repeated indefinitely or until some specified condition is fulfilled. **3** In a system of standing waves, the region separating two NODES.

loose coupling   For two resonant systems, a degree of coupling that is less than critical coupling.

lorac *n.* A system similar to LORAN used mainly for hydrographic and land surveying. The desired position is determined by phase comparison of two continuous wave signals transmitted from widely separated stations on one frequency and the same two signals transmitted from a third station on another frequency.

loran *n.* A long range navigation system, in which phase comparison of radio waves from three fixed stations are used in determining position. A master station transmits pulses at fixed intervals and a SLAVE STATION, 200 to 300 miles away, repeats the pulses with a delay of one half period. The lines of constant phase difference define a hyperbola on the map for the possible position of the navigator. A similar pair of pulses from the master and another slave define another hyperbola, the intersection of the two hyperbolas determining the navigator's position.

Lorentz force A force **F** exerted on an electric charge *q* moving with velocity **v** as a result of the influence of an electric field **E** and a magnetic field **B**, given by $\mathbf{F} = q(\mathbf{E} + \mathbf{v} \times \mathbf{B})$.

loss *n.* Any energy expended uselessly in a system or device, in particular a decrease in signal power in the course of transmission.—**lossy** *adj.*

losser *n.* **1** A dielectric material having appreciable LOSSES. **2** An element included in a circuit to dissipate energy and thus prevent oscillation.

Lossev effect The production of radiation as a result of the recombination of charge carriers (holes and electrons) injected into a forward-biased PN-JUNCTION or PIN-JUNCTION.

lossey *adj.* Of, like, or made of an insulating material capable of dampling out an unwanted mode of oscillation while having little effect on a desired mode.

lossless line An ideal transmission line with no LOSS.

loudness *n.* The subjective attribute of a sound which undergoes the most marked variation as a result of changes in SOUND PRESSURE or INTENSITY.

loudness contour On a graph with SOUND PRESSURE

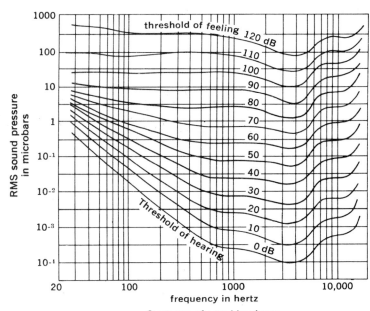

Contours of equal loudness.

level on the ordinate and frequency along the abscissa, any of the curves connecting points for which the sensation of loudness for a listener with average normal hearing is constant.

loudness control A volume control provided with an electrical network that compensates for the variation in frequency response of the ear at different levels of SOUND PRESSURE. See FLETCHER-MUNSEN EFFECT.

loudness level The loudness of a sound as measured in PHONS.

loudspeaker *n.* An electroacoustical transducer for converting electrical signals into equivalent sound signals for general coverage.

low band The band of radio frequencies between 54 and 88 Mhz, including television channels 2 through 6.

low frequency A radio frequency in the band between 30 and 300 khz.

low pass filter A filter that (ideally) passes all frequencies below a certain frequency with negligible attenuation while providing virtually infinite attenuation at frequencies above that.

low tension Low voltage.

LP record LONG PLAY RECORD.

lp.W., L.P.W., L.p.w. *abbr.* Lumen per watt.

lumen *n.* A measure of luminous FLUX defined as equal to the quantity of flow of light through a solid angle of one STERADIAN from a uniform point source of one international candle.

luminance *n.* **1** The luminous intensity of a surface in a given direction; photometric brightness. Measured in candle/cm² or in lamberts. **2** In color television, the degree of brightness, as opposed to the difference in CHROMINANCE.

luminescence *n.* The emission of light by a material due to the absorption of energy at temperatures lower than red heat.

luminophor *n.* A luminescent material that emits light when stimulated by radiation or other forms of energy. Compare PHOSPHOR.

lumped *adj.* Concentrated in single, discrete elements rather than being DISTRIBUTED throughout the fabric of a system.

lutetium (Lu) Element 71; Atomic wt. 174.99; Melting pt. 1652°C; Boiling pt. 3327°C; Density 9.84 g/ml; Valence 3 or 4; Spec. heat 0.037 cal/(g)(°C), Electrical conductivity (0–20°C) 0.015 × 10⁶ mho/cm.

lux *n.* A measure of illumination; 0.0929 foot-candle; 1 lumen/m².

Luxemberg effect CROSS MODULATION caused by the non-linear transmission characteristics of the ionosphere.

M

M *symbol* **1** (*m*) Meter. **2** (*M,m*) Mass. **3** (*m*) Modulation factor. **4** (*m*) Milli-. **5** (*M*) Mega-. **6** (*M*) Magnetization. **7** (*m*) Magnetic moment.

ma *abbr.* Milliampere(s).

mÅ *abbr.* Milliangstrom(s).

macroinstruction *n.* In a digital computer, a type of INSTRUCTION that generates some other set of instructions.

MADT *abbr.* A MICROALLOY DIFFUSED BASE TRANSISTOR.

magamp MAGNETIC AMPLIFIER.

magic eye tube An ELECTRON RAY TUBE.

magic tee HYBRID TEE.

magnesium (Mg) Element 12; Atomic wt. 24.32; Melting pt. 651°C; Boiling pt. 1110°C; Spec. grav. (20°C) 1.74; Hardness 2; Valence 2; Electron arrangement in free atoms (2) (8) 2; Atomic vol. 14.0 cm³/g-atom; Crystal structure Close-packed hexagonal; Spec. heat 0.25 cal/(g)(°C); Electrical conductivity (0–20°C) 0.224 × 10⁶ mho/cm. Used as a GETTER in radio tubes, and on cathodes of phototubes.

magnet *n.* A piece of ferrogmetic material whose magnetism has been so aligned that it produces a magnetic field. It will attract magnetic material and may (depending on orientation) experience torque when placed in another magnetic field.

magnetic amplifier A device in which a control signal

applied to a system of SATURABLE REACTORS modulates the flow of an alternating current in a power curcuit.

magnetic cartridge　In a phonograph, a cartridge that generates audio signals by varying the magnetic flux that links a coil whose terminals are connected to the input of the preamplifier.

magnetic circuit　A closed path of MAGNETIC FLUX having at each point the direction of the MAGNETIC INDUCTION, analogous to an ELECTRIC CIRCUIT.

magnetic constant　PERMEABILITY.

magnetic drum　In a digital computer, a storage device consisting of a rapidly rotating cylinder on which information is stored by magnetized points on the surface.

magnetic field　A region in space in which either of the vectors magnetic induction **B** or magnetic field strength **H** have significant values.

magnetic field strength　A vector **H** that gives the magnitude and direction of the force exerted on a unit MAGNETIC POLE at any point in a magnetic field.

magnetic flux　The number of lines of MAGNETIC INDUCTION passing through a given surface in a magnetic field, given by $\phi = \int_s \mathbf{B} \cdot \mathbf{ds}$, where **ds** is an element of the surface, **B** is magnetic induction, and the indicated integral is taken over the entire surface.

magnetic flux density　MAGNETIC INDUCTION.

magnetic head　A specially designed electromagnet that can be energized by a time-varying current and used for introducing information into MAGNETIC TAPES or DRUMS, and extracting or erasing it.

magnetic induction　A vector **B** associated with the component of mechanical force $d\mathbf{F}$ exerted on a conductor of length dl that carries a current I while at a point in a magnetic field. This force is given by

$$d\mathbf{F} = I d\mathbf{l} \times \mathbf{B}.$$

Alternatively, in a linear, homogeneous, isotropic medium,

$$\mathbf{B} = \mu \mathbf{H},$$

where μ is the MAGNETIC CONSTANT and **H** the MAGNETIC FIELD STRENGTH.

magnetic intensity　MAGNETIC FIELD STRENGTH.

magnetic line of force 1　A line whose tangent at any point represents the direction of MAGNETIC FLUX. **2** A unit of magnetic flux equal to 1 MAXWELL.

magnetic moment　A vector quantity which is a measure of the torque exerted on a magnet, current loop, moving charge, etc., when placed in a magnetic field in which there is a flux density **B**. This torque **T** is given by

$$\mathbf{T} = \mathbf{m} \times \mathbf{B}$$

where **m** is the magnetic moment. Also

$$\mathbf{T} = /\, \mathbf{m}\, \mathbf{B}\, /\, \sin \theta$$

and is a maximum when **m** is perpendicular to **B**.

magnetic pole　A convenient but fictitious entity used in describing various magnetic phenomena. The field of a simple bar magnet, for example, is said to emanate from a "pole" located at one end and return to another "pole" at the other end.

magnetic pole strength　The strength assigned to a MAGNETIC POLE, that is, given a magnet that has a moment **m** and two poles separated by a distance **l**, the pole strength is assigned on the basis of the relation

$$\mathbf{m} = /\, p\, /\, \mathbf{l}.$$

where p, $-p$ are the pole strengths.

magnetic potential　The work per unit magnetic pole spent in moving a magnetized body in a magnetic field from a reference point to a point of interest (P). Commonly the reference point is chosen as infinity. Analytically, assuming a magnetic field strength **H**, the potential (V) is given by

$$V = \int_{\infty}^{P} \mathbf{H} \cdot \mathbf{ds},$$

where **ds** is a vector element of the path from ∞ to P.

magnetics *n.*　The science of MAGNETISM.

magnetic tape　A thin ribbon of plastic or other material coated with a magnetic substance capable of retaining signals fed onto it.

magnetism *n.*　**1** The quality possessed by a magnet or by ferromagnetic material surrounded by an electric current; the specific quality of a magnet regarded as an effect of molecular interaction. **2** The science of magnetic phenomena.

magnetite n.   Ferroferric oxide ($FeO.Fe_2O_3$), a naturally occurring FERRITE.

magneto n.   A small ac generator.

magnetohydrodynamics n.   The study of the behavior of electrically conductive fluids (plasmas in particular) under the influence of magnetic fields. The principal application of this study is in the development of thermonuclear reactors.

magnetoionic adj.   Of or having to do with the effects on the propagation of electromagnetic waves due to the ionization of the atmosphere and the presence of the earth's magnetic field.

magnetoionic wave component   Either of the pair of elliptically polarized components into which a linearly polarized electromagnetic wave is split as it passes into the ionosphere in the presence of the earth's magnetic field.

magnetomotive force   In a MAGNETIC CIRCUIT, a quantity $\mathfrak{F}$ that is the analog of electromotive force, given by

$$\mathfrak{F} = \int \mathbf{H} \cdot d\mathbf{l},$$

where **H** is the MAGNETIC FIELD STRENGTH and l is the length of the magnetic circuit. It can be shown that around a closed contour

$$\mathfrak{F} = 4\pi l,$$

that is

$$\oint \mathbf{H} \cdot d\mathbf{l} = 4\pi l,$$

where l is the current linked by the contour.

magneton n.   BOHR MAGNETON.

magnetostatics n.   The study of magnetic fields that remain invariant with time.

magnetoresistance n.   The variation in the resistivity of a conductor or semiconductor in response to the application of a magnetic field. The effect is especially large in indium antimonide.

magnetoresistor n.   A semiconductor device that exhibits MAGNETORESISTANCE.

magnetostriction n.   A change in the dimensions of a body in a magnetic field. The longitudinal and transverse strictions are usually converse. The effect is used in transducers for the reception and transmission of high-frequency sound.**—magnetostrictive** adj.

magnetron n.   An electron tube, useful for microwave frequencies, in which a beam of electrons interacts with electrostatic and magnetostatic fields that are perpendicular to each other and to the direction of electron flow. If an alternating electromagnetic field is coupled to the electron beam through a suitable resonant cavity the electrons become bunched in such a way that an amplified version of the alternating field can be withdrawn by means of another cavity.

magnet wire   Insulated copper wire used in the coils of an electromagnet.

magnistor n.   A device that uses the magnetoresistive properties of semiconductors such as indium antimonide.

mag-slip   SELSYN.

major apex face   In a quartz crystal, one of the three larger faces adjoining the apex.

major cycle   In a digital computer, the maximum time needed to read any part of a storage device.

major defect   A malfunction causing a complete breakdown in a device.

major face   MAJOR APEX FACE.

major failure   A malfunction in a system which, while not causing complete breakdown, causes it to function uselessly.

majority carrier   In a semiconductor, the type of current carriers that are in the majority, that is, the electrons in an N-TYPE SEMICONDUCTOR or the HOLES in a P-TYPE SEMICONDUCTOR.

majority gate   A logic circuit having an odd number of inputs and a single output which becomes energized whenever a majority of the inputs are energized.

major lobe   See LOBE.

major loop   In a system using FEEDBACK, the LOOP composed of the path from the system input to the system output plus the main feedback path.

make n.   The closing of a contact.

make-before-break   A double throw contact which switches on one circuit shorting it to the other before the latter is switched off.

manganese (Mn)   Element 25; Atomic wt. 54.93; Melting pt. 1260°C; Boiling pt. 1900°C; Spec. grav. (20°C) 7.2; Hardness 5; Valence 2, 3, 4, 6, or 7; Spec. heat 0.115 cal/(g)(°C); Linear coefficient of expansion (α–Mn 20°C) 22.3 $\times$ 10^{-6}; Electrical conductivity (0–20°C) 0.054 $\times$ 10^6 mho/cm. Used as a constituent of various steels and other alloys, in particular, MANGANIN and other high-resistance metals, and also in alloys having a high temperature coefficient of expansion for use in thermostats, etc. Manganese dioxide is used as a depolarizer in dry cells.

Manganin *n.* An alloy of copper, manganese and nickel (84% Cu, 12% Mn, 4% Ni) with a temperature coefficient of resistivity of ±0.00002, used as a standard of resistivity. Melting point 910°C.

Marconi antenna An antenna grounded at one end and radiating from the other.

marker pip A spot or inverted V of light appearing on the surface of a cathode ray oscilloscope to indicate the frequency of a sweep-driven signal generator.

maser *n.* A device analogous to a LASER, designed to operate at microwave frequencies. While the maser is, in fact, the older device, the more varied applications of coherent light have given the laser greater prominence.

masking *n.* The effect of one sound on the audibility or intelligibility of another; the number of decibels the threshold of audibility for a sound is raised in the presence of another sound.

masonite *n.* A board material of compressed steam-exploded wood fibers, often used as a base in electronic devices.

mass *n.* The inertial resistance of a body to acceleration, considered, in classical physics, to be a conserved quantity independent of speed. It can be shown that the parameters m_1, m_2 appearing in the GRAVITATION equation are equivalent to the masses of the bodies in the sense given above. In relativistic physics it is shown that when the speed of a body becomes an appreciable fraction of c, the speed of light mass is given by

$$m = m_0/[1-(v/c)^2]^{1/2},$$

where m_0 is the rest mass of the body and v is its speed relative to the observer who finds its mass to be m. Further, it has been shown that mass and energy are interconvertible as given by Einstein's equation $E = mc^2$.

mass spectroscope An instrument for measuring the mass of atoms and molecules. A sample of the material in gas or vapor form is bombarded by electrons; the ions formed by this bombardment are deflected by magnetic and electric fields onto a photographic plate. The ions form lines on the plate according to their mass and charge.

mast *n.* A pole for supporting an antenna.

master *n.* An element of a system that controls or initiates the action or responses of the other elements of the system.

match *v.* To adjust or arrange (the output and input impedances of two transducers connected in cascade) so that at their junction each impedance is the IMAGE IMPEDANCE of the other.—**matching** *n.* —**matched** *adj.*

matrix *n.* **1** A rectangular array of numbers X_{ij}, where the subscripts *i*, *j* indicate respectively the row and column in which each *X* is located. A matrix in which $i_{max} = r$ and $j_{max} = s$ is called an r by s (r X s) matrix. Matrices have rules for their manipulation and are often valuable in circuit analysis. **2** A network, normally acting as an encoder or decoder, in which a set of input signals is transformed to a different set of output signals.

maximally flat coupling CRITICAL COUPLING.

maximum *n.* The value of a function $f(x)$ at a point x_o where $f(x_o)$ is algebraically the largest value of $f(x)$ in some interval about x_o. If the derivatives $f'(x)$ and $f''(x)$ exist at x_o and $f'(x_o) = 0$, $f''(x_o) < 0$, the function $f(x)$ has a maximum at x_0.

maximum-effort control system A feedback control system which is nonlinear in that the compensating device operates either full "on" or full "off." Systems of this type appear to have advantages in speed of response over linear systems.

maxwell *n.* A cgs unit of magnetic flux equal to 10^{-8} WEBER.

Maxwell bridge or **Maxwell-Wien bridge** An ac BRIDGE used to measure an inductance and any series resistance it may contain respectively in terms of a capacitance and a resistance shunted across the capacitance.

$$L_x = R_a R_b C_3$$

$$R_x = \frac{R_a R_b}{R_s}$$

$$Q_x = \omega \frac{L_x}{R_x} = \omega C_s R_s$$

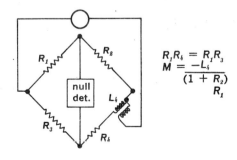

Maxwell inductance-comparison bridge An ac BRIDGE in which the ratio of a pair of inductances located in two adjacent arms is measured in terms of the ratio of the resistances in the other two arms.

Maxwell mutual-inductance bridge An ac bridge in which MUTUAL INDUCTANCE is measured in terms of SELF INDUCTANCE and resistance.

$$R_1 R_4 = R_1 R_3$$
$$M = \frac{-L_4}{(1 + R_2)} R_1$$

Maxwell's equations A set of four partial differential equations that form the basis of electromagnetic theory. They are, in vector notation:

1 Curl $\mathbf{H} = \frac{\partial \mathbf{D}}{\partial t} + \mathbf{J}$

2 Curl $\mathbf{E} = \frac{-\partial \mathbf{B}}{\partial t}$

3 Div $\mathbf{B} = 0$

4 Div $\mathbf{D} = \rho$

where **H** is magnetic field strength, **B** magnetic induction, **E** electric intensity, **D** electric induction, ρ electric charge density, **J** conduction current density, and t time. See CURL; DIVERGENCE.

mayday n. The international radiotelephonic call for immediate help.

mb abbr. Millibar(s).

Mc abbr. **1** Megacycle(s). **2** Megacycles per second.

mC abbr. Millicoulomb(s).

mc abbr. **1** Megacycle(s). **2** Millicurie(s). **3** Millicycle(s).

mcw, MCW abbr. Modulated continuous wave.

m-derived filter section An IMAGE-PARAMETER FILTER section that contains resonant combinations of inductance and capacitance in the series arm, shunt arm, or both. For a section of this kind there is a frequency ω_∞ in the stop band at which the attenuation is infinite. This is given by

$$\omega^2 = \omega_c{}^2 (1 - m^2)$$

where ω_c is the cutoff frequency and m is a design parameter such that $0 < m \leq 1$. The individual reactances X are determined from the prototype CONSTANT-K FILTER section and are given by $X_m = f(m, x_{k,})$ where X_k is the constant-k value.

M-display In radar, a modification of an A-DISPLAY in which a locally generated pulse, located on a PEDESTAL, is displayed along with the target pulse. Distance to the target is measured by moving a calibrated control until the local pulse coincides with the target pulse.

MDI abbr. Magnetic direction indicator.

meaconing n. The rebroadcasting of enemy beacon signals in order to induce inaccurate bearings.

mean life The average period for which an atom or other system remains in a certain state. In particular, in semiconductors, the average time taken by carriers to recombine with others of the opposite sign.

measurand n. A measured physical quantity.

mechanical rectifier   A rectifier using a synchronous mechanical switch, either a VIBRATOR or a rotating wheel, to convert a phased current to dc.

medium *n.*   An ambiance in which a physical force or influence exists. Air, for example, may be a medium for sound; magnetic fields may exist in a non-material medium.

medium frequency   A radio frequency in the band between 300 and 3,000 khz.

meg *n. informal*   MEGOHM.

mega- *prefix*   One million (10^6) times a specified unit: *megahertz; megavolt; megohm.*

megacycle *n.*   One million cycles. Compare MEGAHERTZ.

megahertz *n.*   One million cycles per second.

megatron *n.*   A DISK-SEAL TUBE.

Megger *n.*   An instrument for measuring high resistivity, using a built-in hand-cranked dc generator: a trade name.

megohm *n.*   One million ohms.

Meissner effect   The essentially perfect DIAMAGNETISM displayed by certain materials when they are placed in a magnetic field and cooled to temperatures at which they assume SUPERCONDUCTIVITY. In effect the lines of magnetic induction are forced out of the material.

Meissner oscillator   An oscillator in which the feedback path consists of a resonant LC circuit to which the input and output circuits are inductively coupled.

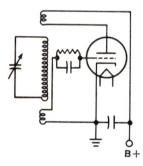

mel *n.*   The measure of PITCH, defined as 0.001 of the difference in pitch between a pure tone of 1000 hz (which is by definition 1000 mels) and another pure tone that the listener judges to be twice as high.

M.El.Eng. *abbr.*   Master of Electrical Engineering.

meltback process   A method of making semiconductor junctions by melting a properly doped crystal and allowing it to recrystallize.

memistor *n.*   A storage device in a computer which operates by the electrolytic deposit or removal of copper on or from a resistive substrate. The variation in the substrate's resistivity provides the signal.

memory *n.*   In a digital computer, STORAGE.

mendelevium (Md)   Element 101. Radioactive element with half-life of 3 hours; decays by spontaneous fission.

mercury (Hg; *Lat* hydragyrum**)**   Element 80; Atomic wt. 200.61; Melting pt. −38.87°C; Boiling pt. 356.9°C; Spec. grav. (20°C) 13.546; Valence 1 or 2; Thermal expansion 0–100°C approximately that of a gas; Spec. heat 0.033 cal/(g)(°C); Surface tension (liquid) 505 dyne/cm; Thermal conductivity (0°C metal) 2.2% that of silver. Electrical conductivity (0°C metal) 1.58% that of silver. Used as a liquid contact for electrical switches, and for the manufacture of mercury vapor lamps and rectifiers; also used in electrochemistry for the standard calomel electrode, as the reference electrode for the measurement of potentials and for potentiometric titrations, and for the Weston standard cell. Its alloys and compounds are diamagnetic.

mercury arc tube   A discharge tube, rectifier tube, etc. filled with mercury vapor.

mercury cell   A dry cell having a cathode of mercuric oxide (HgO) mixed with graphite, an anode of zinc, and an electrolyte of potassium hydroxide (KOH) saturated with zinc oxide (ZnO).

mercury switch   An electric switch using liquid mercury to make and break contacts.

mercury vapor rectifier   A rectifier tube in which the conduction is by arc discharge through mercury vapor, with a resulting low voltage drop across the tube.

mesa 1 *n.*   In certain transistors, a raised area (resembling somewhat a geological mesa) left when semiconductor material is etched away to allow access to the base and possibly, collector, regions. **2** *v.* To make a (transistor, junction, etc.) by a process that involves the formation of a mesa.

mesa transistor A transistor in which access to one or more regions is obtained by an etching process that leaves a MESA.

mesh *n.* In an electrical NETWORK, a LOOP that does not encircle or enclose another loop and which cannot be divided into other loops.

meson *n.* Any elementary particle whose rest mass is more than that of an electron and less than that of a proton.

meta- *prefix* Changed; behind; later; with.

metadyne *n.* AMPLIDYNE.

metal *n.* Any of the elements that in general form cations, characterized by generally high thermal and electrical conductivity, luster, ductility, malleability, etc.—**metallic** *adj.*

metallic circuit An ungrounded circuit.

metallized capacitor A capacitor in which a film of metal is deposited on the dielectric. Any failure in the dielectric involves the collapse of neighboring portions of the film and the capacitor is thus self-healing.

metalloid *n.* An element, as arsenic, antimony, etc., whose physical properties are generally metallic but whose chemical properties resemble those of a metal in some cases and those of a nonmetal in others.

metal-oxide semiconductor field-effect transistor A FIELD-EFFECT TRANSISTOR in which the GATE is insulated from the CHANNEL by a metal-oxide dielectric. This allows the gate to be forward-biased in order to enhance channel conductivity and makes the input resistance of the device extremely high.

metastable *adj.* Indicating an energy state which, while essentially unstable, does not change without external stimulation.

meter *n.* The fundamental unit of length in the metric system, now defined as 1,553,164.13 wavelengths of the red light of a cadmium emission spectrum in air at 730 mm pressure at 0°C.

meter relay A relay which trips when a measuring device records a certain quantity.

metrechon *n.* A type of electrostatically-operated STORAGE TUBE.

metric system A decimal measuring system based on the GRAM and the METER.

mf *abbr.* **1** Medium frequency. **2** Millifarad. **3** Microfarad.

MF *abbr.* Middle frequency.

mfd. *abbr.* Microfarad.

m.g.s. *abbr.* Meter-gram-second.

mh *abbr.* Millihenry.

MHD *abbr.* Magnetohydrodynamics.

mho *n.* A measure of conductance, the reciprocal of the OHM.

mica *n.* A silicate that tends to cleave into sheets having good heat resistance and insulation. Its dielectric constant is 5.4, dielectric strength between 118 and 276 volts/mil at 25°C with a thickness of 0.04 in., and resistivity 5×10^{13} ohm/cm.

micro- *combining form* **1** One millionth (10^{-6}) of a specified quantity or dimension. **2** Small.

microalloy transistor A transistor which is made by etching depressions into opposite faces of a blank of semiconductor material of one type which eventually forms the base. Donor contacts of the opposite types are electrolytically deposited in these depressions and subsequently heat treated and converted to alloy contacts so as to form the emitter and collector regions.

microalloy diffused transistor A MICROALLOY TRANSISTOR in which the impurity content of the original semiconductor blank is introduced by diffusion, thus allowing a nonlinear impurity profile in the base region.

microgroove *n.* The groove on a LONG PLAY RECORD, usually about 0.001 in. wide.

microbar *n.* One dyne/cm².

micromicro- PICO-.

micron *n.* 10^{-6} meter.

microphone *n.* A transducer for converting sound waves into equivalent electrical signals. See TRANSDUCER; CAPACITOR MICROPHONE, CARBON MICROPHONE, CONTACT MICROPHONE, CRYSTAL MICROPHONE, HOT WIRE INSTRUMENT, LAVALIER MICROPHONE, LINE MICROPHONE, PRESSURE MICROPHONE, RIBBON MICROPHONE.

microphonics *n.* Noise caused by the vibration of the elements of a tube in an electronic system. While usually undesirable, microphonics can sometimes be used to detect strains.

microvolts per meter   A measure of the intensity of a radio signal at a certain point, defined as the intensity in microvolts of the signal induced in a 1 meter length of wire held in an rf field so that it is perpendicular to the direction of wave propagation and parallel to the direction of wave polarization.

microwave *n.*   A wave occupying a position in the wave spectrum between the infrared and conventional radio frequencies from about 1,000 to 300,000 Mhz with wavelengths of from 30 cm to 1 mm.

midband frequency   A CENTER FREQUENCY.

mid-range *n.*   The band of audio frequencies from roughly 600 to 6000 hz.

mike *n. informal*   MICROPHONE.

mil *n.*   **1** 0.001 in. **2** A unit of angular measurement used in guidance systems defined as an arc of 0.001 radian.

Miller bridge   An ac BRIDGE circuit used for the measurement of the AMPLIFICATION FACTOR μ or TRANSCONDUCTANCE g_m of vacuum tubes.

Miller effect   An effect whereby the input capacitance of a triode amplifier is effectively the sum of grid-cathode capacitance plus roughly the gain of the tube multiplied by the grid-plate capacitance; analytically,

$$C_{in} = C_{gk} + (1 + A)C_{gp}$$

where A is the gain of the stage and the subscripts denote the respective capacitances.

Miller integrator   An integrator circuit consisting of an operational amplifier with a resistor in series with its input and a capacitor in its feedback loop.

Miller oscillator   A crystal-controlled oscillator in which the frequency of operation is the parallel resonant frequency of the crystal, the active element in the circuit acting as a NEGATIVE RESISTANCE connected in parallel with the crystal.

milli- *Combining form*   One thousandth (10^{-3}) of a specified quantity or dimension.

millipercent *n.*   0.001%.

miniature tube   A small electron tube with a 7 or 9 pin base.

minimum *n.*   The value of a function $f(x)$ at a point x_0 where $f(x_0)$ is algebraically the smallest value of $f(x)$ in some interval about x_0. If the derivatives $f'(x)$ and $f''(x)$ exist at x_0 and $f'(x_0) = 0$, $f''(x_0) > 0$, the function $f(x)$ has a minimum at x_0.

minitrack *n.*   A method of tracking satellites by phase comparison of signals from the satellite as it passes over a minitrack station's base line. This fixes the satellite's angular position from the station.

minor lobe   See LOBE.

minority carrier   In a semiconductor, current carriers of the type that are in the minority, that is, the electrons in a P-TYPE SEMICONDUCTOR or the HOLES in an N-TYPE SEMICONDUCTOR.

Minter stereo system   A proposed system of stereo recording that is designed to be compatible with monophonic equipment. The system is basically similar to the CROSBY SYSTEM of FM stereo broadcasting except that the carrier for the L-R information is at 25 khz and frequency modulation is used.

minute *n.*   **1** A measure of angle, equal to 1/60 degree. **2** A measure of time, equal to 60 seconds.

misfire *n.*   In a mercury pool tube, failure to establish an arc between cathode and anode in the required period.

mistake *n.*   In computer work, an incorrect action, as faulty programming or bad arithmetic, performed by a human being. In this field, a distinction is often made between a *mistake* and an ERROR.

m.k.s., mks, MKS, M.K.S. *abbr.*   Meter-kilogram-second.

mix *v.*   To combine (two or more input signals) in a TRANSDUCER so as to produce a single output. In general, two cases arise. In the first case the transducer is LINEAR and the output consists of a SUPERPOSITION of the input signals. In the second case the transducer is nonlinear and the output consists of the HETERODYNE products of the input signals.—**mixer** *n.*

mm *abbr.*   Millimeter(s).

M.M.E. *abbr.*   Master of Mechanical Engineering.

mmf, MMF *abbr.*   **1** Micromicrofarad. **2** Magnetomotive force.

m.m.s. *Br. abbr.*   Millimeter-milligram-second.

mobility *n.*   The average velocity of carriers in a semiconductor per unit of the applied electric field. It is often measured by the HALL EFFECT and is given by

$$\mu = \sigma R_H$$

where μ is the mobility, R_H the Hall coefficient, and σ the conductivity.

mode *n.* **1** Any of the characteristic ways in which an oscillatory system can respond. In general, the modes of a system correspond to the solutions of the DIFFERENTIAL EQUATION or equations that describe its behavior. Alternatively, the modes may be represented as the POLES of a frequency response function. **2** Of a wave, the disposition of its field variations with respect to its direction of propagation. In the **transverse electric mode** (TE) the electric vector is confined to directions perpendicular to the direction of propagation, in the **transverse magnetic mode** (TM) the magnetic vector is confined to these directions; in the **transverse electromagnetic mode** (TEM) both of these vectors are so confined. In waveguides the TE and TM modes are given a pair of subscripts, the first indicating the number of half-period field variations along the diameters of a circular waveguide or along the larger transverse axis of a rectangular waveguide, the second indicating the number along the radii of a circular guide or along the smaller transverse axis of a rectangular guide. The TEM mode cannot propagate in a hollow waveguide.

modular *adj.* Of, having to do with, or constructed from modules.

modulate *v.* To vary some characteristic of (an electromagnetic wave, electric current, etc.), generally in order to impress on it information that is to be transmitted. Assuming, for example, a sinusoidal carrier of the form

$$C(t) = A \cos(\omega t + \phi),$$

modulation could be accomplished by varying A, ω, or ϕ, or several of these at once, in accordance with some modulating function.—**modulator** *n.*—**modulation** *n.*

module *n.* A complete subassembly of a larger system combined in a single package.

moire *n.* An unwanted pattern in a television picture tube caused by interference between two periodic effects in the image.

molectronics or **moletronics** *n.* MOLECULAR ELECTRONICS.

molecular electronics A part of electronics concerned with the development of microcircuitry in which materials, generally semiconductors, are modified so that while each module as a whole acts as a particular equivalent circuit it is not possible to identify discrete elements within the module.

molecule *n.* The smallest unit into which a substance may be divided without changing its chemical properties.

molybdenum (Mo) Element 42; Atomic wt. 95.95; Melting pt. 2620°C; Boiling pt. 5560°C; Spec. grav. 10.2; Valence 2, 3, 4, 5, or 6; Spec. heat 0.061 cal/(g)(°C); Electrical conductivity (0–20°C) 0.19 $\times$ 10^6 mho/cm. Its compounds are used in electroplating baths to give black coatings for decorative and protective purposes.

moment *n.* The product of a quantity and the displacement of the quantity from some significant point about which it acts. See MOMENT OF INERTIA; TORQUE.

moment of inertia The degree to which a body resists ANGULAR ACCELERATION about a given axis. For a rigid body of uniform density ρ the moment of inertia with respect to the x-axis (I_X) is given by

$$I_X = \int (y^2 + z^2)\, dm = \int (y^2 + z^2)\, \rho\, dV,$$

where dm is the element of mass contained in the infinitesimal element of volume dV, and the limits of integration depend on the extent of the body.

momentum *n.* A quantity defined as the product of the mass (m) and velocity of a body. In vector notation momentum (ρ) is given by $\rho = m\mathbf{v}$. The relation between force $(\mathbf{F})$ and momentum is

$$\mathbf{F} = d\rho/dt = m\, d\mathbf{v}/dt + \mathbf{v}\, dm/dt.$$

monaural *adj.* Describing a sound reproduction system with one information channel.

monitor *n.* **1** A device for measuring and recording the performance of an instrument or system, particularly one used in radio communication. **2** The operator of a television monitoring system who chooses one out of several camera images for broadcasting.

monkey chatter A garbled sound produced when a signal is interfered with by the sidebands of an adjacent channel.

monochromatic *adj.* **1** Having only one color, or color with a very narrow band of frequencies. **2** Describing any radiation with an extremely narrow range of frequencies or with every particle having the same or nearly similar energy.

monochromator *n.* A device, such as a LASER, that produces monochromatic light.

monochrome *adj.* In television, black and white. The monochrome channel carries LUMINANCE information in color television.

monolithic circuit INTEGRATED CIRCUIT.

monophonic *adj.* MONAURAL.

monopole antenna A vertical antenna consisting essentially of half a dipole, that is, a conductor that is 0.25 wavelength at the operating frequency with a voltage node at the lower end and a current node at the top.

monopulse radar Radar in which high-precision directional information is obtained by using an array of receiving antennas whose patterns have two or more partially overlapping lobes, sum and difference signals or phase comparison signals being formed from the antenna outputs.

monoscope *n.* An electron tube in which a video test signal is generated by scanning an electron beam across an anode that has secondary emission characteristics that vary from point to point according to the pattern to be produced.

monostable *adj.* Denoting a circuit that has one permanently stable state and one quasi-stable state. When triggered it undergoes rapid transition from the stable state to the quasi-stable state, remains there for a time and spontaneously returns to the stable state.

monostable multivibrator A MULTIVIBRATOR which is stable in only one of its two states.

monostatic *adj.* Describing a radar system in which transmitter and receiver are located together.

morphological circuitry Circuitry whose operation and characteristics depend on the molecular structure of crystals. See MOLECULAR ELECTRONICS.

Morse code A system of telegraphic symbols in which letters and digits are represented by dots and dashes.

mosaic *n.* A mica sheet covered with spots of cesium-silver oxide used in an ICONOSCOPE. The spots retain a positive charge when bombarded with light and absorb electrons from a scanning beam in proportion to the amount of light they have received.

mosfet A METAL OXIDE SEMICONDUCTOR FIELD EFFECT TRANSISTOR.

motional impedance In an electromechanical transducer, the difference between its impedance when normally loaded and its BLOCKED IMPEDANCE.

motor *n.* Any transducer that converts electrical energy to mechanical energy.

motor board The mechanical parts of a tape recorder, or the platform on which they are mounted.

motorboating *n.* In an audio system, unwanted oscillation at a subsonic or low audio frequency.

mountain effect The reflection of radio waves from irregularities in terrain. It causes errors in radio direction finding signals.

mouth *n.* The larger end of a horn.

moving coil loudspeaker A loudspeaker in which the diaphragm is attached to a coil situated in the field of a permanent magnet. Variations in current through the coil cause it to move in reaction to the field.

MP *abbr.* Mounting panel.

MPD *abbr.* Maximum permissible dose.

MPE *abbr.* Maximum permissible exposure.

mR, mr *abbr.* Milliroentgen(s).

mrem *abbr.* Millirem.

mS *abbr.* Millisiemens.

M scan M DISPLAY.

msec *abbr.* Millisecond(s).

MTBF *abbr.* Mean-time-between-failures.

MTI *abbr.* Moving-target indicator.

MTTF *abbr.* Mean-time-to-failure.

MTTFF *abbr.* Mean-time-to-first-failure.

mu (written M, μ) The twelfth letter of the Greek alphabet, used symbolically to represent any of various coefficients, constants, etc., of which in electronics the principal ones are: **1** (μ) AMPLIFICATION FACTOR. **2** (μ) PERMEABILITY. **3** (μ) MICRO-. **4** (μ) MICRON.

MUF *abbr.* Maximum usable frequency.

Muller tube or **Müller tube**   A thermionic vacuum tube having an additional element, usually a grid or auxiliary cathode, connected internally to the main cathode through a large resistor.

multi- *prefix*   Much; many, consisting of many.

multi *n.*   A MULTIVIBRATOR.

multiar *n.*   A comparator that operates by using a diode that is reverse biased by a reference voltage as a series switching element in a regenerative feedback configuration.

multimeter *n.*   VOLT-OHM MILLIAMMETER.

multipactor *n.*   A microwave switching device which functions by using a radio-frequency electric field to drive a thin electron cloud back and forth between a pair of parallel plane surfaces.

multipath *n.*   The reception of a single radio signal through two or more paths, usually due to reflection; a cause of television ghosts.

multiphase oscillator   An oscillator system that delivers several usually sinusoidal outputs that are identical except for a constant phase difference between them.

multiplex *n.*   The transmission of two or more signals over the same channel.

multiplier *n.*   **1** An electrical network whose output is proportional to the product of its inputs. **2** A circuit element that can be switched in or out of a voltage- or current-division network, frequency determining network, etc., thus establishing the range of operation of some device. **3** An ELECTRON MULTIPLIER. **4** A FREQUENCY MULTIPLIER.

multiturn potentiometer   A HELICAL POTENTIOMETER.

multivibrator *n.*   A relaxation oscillator consisting of two active elements (vacuum tubes, transistors, etc.) regeneratively coupled in such a way that when one is in a conducting state the other is cut off. Some types make the transition between states spontaneously, while others require triggering. See ASTABLE, DISTABLE, MONOSTABLE MULTIVIBRATOR.

Murray loop test   A method of locating accidental grounds on transmission lines by connecting the suspect wire and a wire known to be intact as two adjacent arms of a Wheatstone bridge.

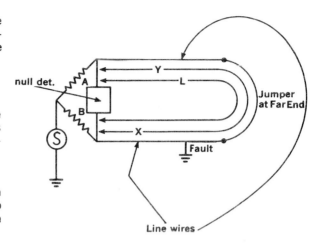

music power   A parameter used in rating the output capability of high-fidelity audio amplifiers. Its value depends upon transient response, power supply regulation, and several other factors as well as rms power capability. In general, the music power rating of an amplifier exceeds its rms power rating.

mutual conductance   TRANSCONDUCTANCE.

mv *abbr.*   Millivolt(s).

MV, Mv *abbr.*   Megavolt(s).

MVA *abbr.*   Megavolt-ampere(s).

mv/m *abbr.*   Millivolt(s) per meter.

mW, mw *abbr.*   Milliwatt(s).

Mw *abbr.*   Megawatt(s).

mx *abbr.*   Maxwell(s).

N

na *abbr.*   Nanoampere(s).

NAB *abbr.*   National Association of Broadcasters.

NAND-gate   A LOGIC CIRCUIT having two or more inputs and a single output, which may be formed by putting an inverter after an AND-GATE. A signal appears at its output unless all of its inputs are energized.

nano- *combining form* One billionth (10^{-9}) of a specified quantity or dimension.

nanocircuit *n.* A flipflop or other logic circuit which can switch states in less than a nanosecond (10^{-9} second); a combination of such circuits operating in the nanosecond range.

napier *n.* A NEPER.

Napierian logarithm NATURAL LOGARITHM.

NASA *abbr.* National Aeronautics and Space Administration.

natural frequency The frequency or frequencies with which a resonant system oscillates in the absence of damping and/or externally supplied energy. In radians per second the natural frequency ω_o is given by

$$\omega_o = 1/\sqrt{LC},$$

where L is inductance (or one of its analogs) and C is capacitance (or one of its analogs). More generally, a system will have a natural frequency corresponding to each of its NATURAL MODES.

natural logarithm A logarithm using

$$e = \lim_{n\to\infty}\left(1 + \frac{1}{n}\right)^n = 2.71828\cdots$$

as its base.

natural modes The MODES that a resonant system has in the absence of damping or externally supplied energy.

Navaglobe *n.* A long-distance low-frequency radionavigational system using three fixed antennas in the form of a triangle with sides of about 0.4 wavelength. Each pair of antennas in turn transmits for 0.25 second, the three signals determining the aircraft's bearing from the transmitter. A fourth pulse differing by 100hz is transmitted by all three antennas for synchronization. See NAVARHO.

Navarho *n.* A radionavigational system of distance measurement used in association with NAVAGLOBE. A 200hz oscillator with a stability of $1:10^9$ is carried in an aircraft and checked against two simultaneous Navaglobe signals before flight. The phase difference between the ground and air signals measures their distance to an accuracy of $\pm$ 3 miles. Ambiguity between phase differences of multiples of 360° is avoided by use of the fourth Navaglobe pulse.

NC **1** No connection (used of vacuum tube bases). **2** Normally closed (of switches and relays).

N-display In radar, a display combining K-DISPLAY and M-DISPLAY, in which each waveshape is duplicated and a locally generated pulse is located on a pedestal.

near infrared See INFRARED.

needle *n.* STYLUS.

needle talk Unwanted sounds produced directly by a phonograph stylus.

negative feedback See FEEDBACK.

negative resistance A circuit element in which the change in current I as a result of a change in applied voltage E is negative. More explicitly, in a normal resistance R,

$$I = E/R$$

and

$$dI/dE = 1/R;$$

but if dI/dE has a negative value, say $-a$, the element across which this occurs appears, so far as the dynamic conditions are concerned, to have a resistance of $-a$. This behavior is found in tetrodes and tunnel diodes over a part of their voltage-current characteristics and is of great use in oscillator circuits.

negatron *n.* See ELECTRON.

NEMA *abbr.* National Electrical Manufacturers Association.

nemo *n.* A radio or television program broadcast from outside the studio.

neodymium (Nd) Element 60; Atomic wt. 144.27. Melting pt. 840°C; Spec. grav. 6.95; Valence 3; Spec. heat 0.045 cal/(g)(°C); Electrical conductivity (0–20°C) 0.013×10^6 mho/cm.

neon (Ne) Element 10; Atomic wt. 20.183; Melting pt. $-248.67°C$; Boiling pt. $-245.9°C$; Density (0°C) 0.8990 g/1; Valence 0; Liquid density ($-245.9°C$) 1.2 g/ml. Used in filling of neon signs, as a current-carrying agent in lightning arresters, in electron tubes, sparkplug test lamps, and in warning indicators on high-voltage electrical lines.

neper n. A measure used to compare two scalar qualities, as currents and voltages, based on the natural logarithm (ln). When currents I_1, I_2 (or voltages E_1, E_2) are compared, for instance, the number of nepers (N) is given by $N = \ln(I_1/I_2)$ or $N = \ln(E_1/E_2)$. When power values (based on the square of the current or voltage) are compared N is given by $N = \frac{1}{2}\ln(I_1/I_2) = \frac{1}{2}\ln(E_1/E_2)$. One neper = 8.686 decibels.

neptunium (Np) Element 93; Atomic wt. 239; Valence 4, 5, or 6. Radioactive with a half-life of 2.3 days.

Nernst bridge An ac BRIDGE in which all four impedances are capacitances, used for making capacitance measurements at high frequencies.

net n. A number of stations equipped to communicate with each other, often on a common channel and with a definite schedule.

network n. **1** Any interconnection of electrical or electronic devices. **2** A set of interconnected radio or television broadcasting facilities.

network synthesis The branch of electronics concerned with the design of filters and other networks having certain desired properties.

neuristor n. A device that is essentially an active transmission line designed to propagate signals without attenuation. It is an electrical analog of a nerve fiber.

neuroelectricity n. Electricity generated in a living nervous system.

neutral adj. Having a net electric charge of zero.

neutralize v. To connect (an active device such as a transistor or vacuum tube) in such a way that the reactive component of its internal feedback is cancelled. See UNILATERALIZE. **—neutralization** n.

neutrodyne n. A form of tuned radio-frequency amplifier stage in which neutralization is accomplished by connecting a capacitor between the plate and grid circuits.

neutron n. A particle having a REST MASS of 1.6718×10^{-24} g. no charge, a SPIN of one half unit, that is $h/2$ or $h/4\pi$, where h is PLANCK'S CONSTANT, and magnetic moment -1.9125 Bohr magnetons. It forms part of the nucleus of the atom.

newton n. In the mks system, the measure of force, defined as equal to the force needed to accelerate a mass of one kilogram by one meter per second per second.

nF, nf abbr. Nanofarad(s).

nH, nh abbr. Nanohenry(s).

nicad n. A sealed storage battery having a nickel anode, a cadmium cathode and a potassium hydroxide electrolyte.

Nicalloy n. An alloy of 53% iron and 47% nickel having high permeability.

Nichrome n. An electrically resistant alloy of 80% nickel and 20% chromium having resistance to oxidation up to 2100°C., used as a heating element. Nichrome I has about 25% of iron in its composition; Nichrome V has small quantities of manganese, silicon and carbon.

nickel (Ni) Element 28; Atomic wt. 58.69; Melting pt. 1445°C; Boiling pt. 2900°C, Spec. grav. (20°C) 8.90, Valence 2 or 3; Spec. heat 0.105 cal/(g)(°C); Heat conductivity 15% that of silver; Electrical conductivity 15% that of copper, and 14% that of silver. Used as a protective and ornamental coating for less resistant metals, and as a constituent of many alloys designed for special properties, as high resistivity, retention of strength at high temperatures, high magnetic retentivity, etc.

Nilvar n. An alloy similar to INVAR but containing no carbon. It is used for the same purposes.

Nimonic n. INCONEL.

niobium (Nb) Element 41; Atomic wt. 92.91; Melting pt. 2500°C; Boiling pt. 3300°C; Spec. grav. (20°C) 8.4; Valence-electron configuration $4d^46s^1$; Spec. heat 0.065 cal/(g)(°C); Electrical conductivity (0–20°C) 0.080×10^6 mho/cm. Has compounds and alloys that become superconductive at relatively high temperatures.

Ni-Span A series of alloys designed for particular thermal expansion qualities. **Ni-Span-Lo** (45%Fe, 52%Ni, 2.4%Ti, 0.6%Al) has a low coefficient of expansion, **Ni-Span-Hi** (59.6%Fe, 9%Cr, 29%Ni, 2.4%Ti) has a high coefficient, and **Ni-Span-C** (49.5%Fe, 5.5%Cr, 42%Ni, 2.4%Ti, 0.6%Al) has a constant modulus of electricity.

nit n. A measure of LUMINANCE, defined as equal to one CANDLE per square meter.

nitrogen (N)   Element 7; Atomic wt. 14.008; Melting pt. $-209.86°C$; Boiling pt. $-195.8°C$; Spec. grav. ($-195.8°C$ liquid) 0.808, ($-252.0°C$ solid) 1.026; Density 1.2506 g/1; Valence 3 or 5; Heat of transformation ($\alpha\rightarrow\beta$) 54.71 cal/mole; Heat of fusion 172.3 cal/mole; Heat of vaporization 1332.9 cal/mole; Critical temp. $126.26\pm 0.04°K$; Critical pressure 33.54 ± 0.02 atm; Density ($-252.6°C$ α-form) 1.0265 g/ml, ($-210°C$ β-form) 0.8792 g/ml, (liquid) 1.1607 $- 0.0045T$ (T=abs. temp.); Spec. heat 0.247 cal/(g) (°C). Often used as the atmosphere of incandescent light bulbs.

NMR *abbr.*   Nuclear magnetic resonance.

NO *abbr.*   Normally open.

nobelium (No)   Element 102. Radioactive with a half-life 10–22 minutes; emits alpha particles.

noble *adj.*   Chemically inert.

noctovision *n.*   A television system using a camera that is sensitive to infrared and which thus operates in darkness.

nodal diagram   A diagram in which the modes of propagation and wave orders for a waveguide are shown.

node *n.*   **1** In an electrical network, a point that is common to two or more circuit elements. **2** In a system of standing waves, the points, lines, or surfaces where or along which a parameter represented by the wave has an amplitude of essentially zero. See PARTIAL NODE.

noise *n.*   Unwanted currents or voltages in an electrical or electronic system.

noise figure   A figure of merit F for a transmission system, etc., given by

$$F = \frac{S_{in}/N_{in}}{S_{out}/N_{out}}$$

where S_{in} and N_{in} are the signal and noise power at the input and S_{out} and N_{out} are the signal and noise output powers respectively. Both noise powers are referred to a standard temperature of 290°K. Note that $F > 1$ since $N_{out} = GN_{in} + N_{syst}$ and $S_{out} = GS_{in}$ where G is the system power gain and N_{syst} represents the inherent noise power of the system. Thus a low noise amplifier is characterized by a noise figure close to unity.

noise silencer   A LIMITER that silences all noise stronger than the strongest desired signal, reducing the effects of atmospheric and man made noise.

noise temperature   A measure of the internally generated noise of an amplifier, defined as the temperature T_n of a passive system having an available noise power per unit bandwidth which would produce the thermal noise equal to that actually at the system's terminals. Analytically,

$$T_n = 290(F - 1)°K.$$

where F is the **noise function** (a function describing the internally generated noise of the amplifier).

nominal bandwidth   In a BANDPASS FILTER, the difference between the nominal upper and lower cutoff frequencies.

nomogram *n.*   A graphic representation of numerical relations, in particular, a chart typically having scales for three related variables so arranged that a straight line joining values of two of the variables will cut the third scale at the related value of the third variable.

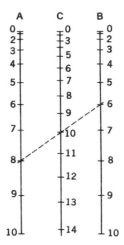

non-polarized electrolytic capacitor   See ELECTROLYTIC CAPACITOR.

normal *n., adj.*   In mathematics, perpendicular.

normal glow   In a gas discharge tube, the glow produced when the tube operates in the region of current-voltage characteristics in which the voltage remains very nearly constant as the current changes.

north pole In a magnet, the pole through which the magnetic lines of force are considered to leave.

Norton's theorem A theorem stating that in a linear, bilateral network all of whose generators operate at the same frequency, the current through an impedance Z_L connected to a pair of terminals a,b is the same as if Z_L were connected to an ideal current generator whose output current equals the short-circuit current between a and b and shunted by an output impedance Z_0 equal to the impedance of the network as seen from terminals a,b, with all independent generators replaced by their internal impedances. If the original network contained dependent sources the output impedance can be found by the ratio of the voltage V_{ab} (with Z_L removed), to the short circuit current i.e., $Z_o = V_{ab}/I_o$.

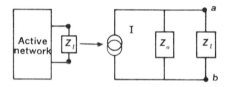

notch *n.* On a graph (a graph of frequency response in particular) a point where a curve dips sharply and returns equally sharply to its original value.

NOT-gate A LOGIC CIRCUIT having one input and one output signal. The output is energized if the input is not, and is not if the input is.

noval base The arrangement of the pins of a nine-pin, miniature, glass vacuum tube.

novice license A two-year, nonrenewable license issuable to ham radio operators who are able to send and receive code at the rate of 5 words per minute and who are familiar with basic radio theory and regulations. The novice must, in addition to observing restrictions on operating frequencies and power input to the final amplifier stage, qualify for a higher class of license at the end of two years.

npn transistor A bipolar transistor in which the emitter and collector are n-type semiconductor material and the base is p-type.

nsec *abbr.* Nanosecond(s).

N.T.P. *British abbr.* Normal temperature and pressure.

N.T.S.C. *abbr.* National Television System Committee.

n-type semiconductor A semiconductor in which due to the presence of DONOR IMPURITIES the MAJORITY CARRIERS are negative electrons.

nu (*written* N, ν) The thirteenth letter of the Greek alphabet, used symbolically to represent any of various coefficients, constants, etc., of which in electronics the principal ones are: **1** (ν) FREQUENCY. **2** (ν) The reciprocal of dispersive power. **3** (ν) RELUCTIVITY. **4** (ν) Wave number.

nucleonics *n.* The branch of science and engineering devoted to the practical application of nuclear energy.

nuvistor *n.* A type of electron tube in which the elements are essentially coaxial cylinders contained in a ceramic envelope.

Nyquist diagram A plot on the complex plane of the negative of the loop gain of a FEEDBACK system, used in evaluating the stability of the system. Given a system in which the transfer function of the forward path is G(s) and that of the feedback path is H(s), a plot is made of the real versus the imaginary part of the product

$$A(s) = G(s)H(s),$$

as s increases from o toward $+j\infty$. According to the **Nyquist stability criterion** the system is stable if and only if the number of counterclockwise encirclements of the point $- 1, + j0$ equals the number of POLES in the right half of the plane. If the system is open-loop stable, closed-loop stability is indicated if the plot does not pass through or enclose the point $-1 +j0$.

Nyquist interval The interval of time occupied by each code element when a communications channel is being used at the NYQUIST RATE.

Nyquist rate The maximum rate N at which code elements can be resolved without ambiguity in a communications channel of a certain limited bandwidth in which the peak noise level is less than one-half the quantum step between code signals, given by

$$N = 2B,$$

where B is the bandwidth of the channel in hertz.

O

o/c *abbr.* Overcharge.

occlude *v.* To take up and hold (a gas) without undergoing any change of chemical properties, as various materials, metals in particular.

octal *adj.* **1** Of or pertaining to eight. **2** Of an electron tube, having eight pins.

octave *n.* The interval between two frequencies whose ratio is 2:1.

OD *abbr.* Outside diameter.

odd function See FUNCTION.

odograph *n.* An electronic device used in a vehicle to trace its course on a map.

oe *abbr.* Oersted.

oersted *n.* The cgs measure of magnetic field strength, defined as the field produced at the center of a plane circular coil of one turn and radius one centimeter carrying a current of $\pi/2$ abamp.

offset *n.* In a process control system, the difference between the desired control point and that actually achieved.

OFHC *abbr.* Oxygen-free, high conductivity.

ohm *n.* A measure of RESISTANCE defined as being equal to the resistance of a conductor carrying a current of one ampere at a potential difference of one volt between the terminals.

ohmic *adj.* Passing a current that is proportional to the applied voltage.

ohmmeter *n.* A device for measuring resistance.

Ohm's law A law stating that the electric current I in a circuit is proportional to the voltage E and inversely proportional to the resistance R. Analytically, $I = E/R$.

Ohm's law for magnetic circuits The statement, analogous in form to OHM'S LAW, giving the relation between MAGNETIC FLUX ϕ, RELUCTANCE $\mathcal{R}$, and MAGNETO-MOTIVE FORCE $\mathcal{F}$, NAMELY that

$$\phi = \mathcal{F}/\mathcal{R}.$$

ohms-per-volt A measure of sensitivity in voltage measuring devices defined by the resistance of the device for a particular range divided by the full-scale voltage indication in that range.

omega (written Ω, ω) The twenty-fourth letter of the Greek alphabet, used symbolically to represent any of various coefficients, constants, etc., of which in electronics the principal ones are: **1** (ω) ANGULAR FREQENCY. **2** (ω) Dispersive power. **3** (ω) Periodicity. **4** (Ω) Volume of phase space. **5** (Ω) OHM.

omni- *prefix* All; totally.

omnibearing *n.* The azimuth as calculated from the signal of an OMNIDIRECTIONAL radio navigation transmitter.

omnidirectional *adj.* **1** Radiating and receiving equally well in all directions. **2** Of an antenna, receiving or radiating equally well in all horizontal directions.

open *n.* A circuit interruption.

open circuit A circuit with an OPEN.

operating point The point on its characteristic curve at which an active element, as a vacuum tube, transistor, etc., operates in the absence of an input signal. The location of this point is determined by the bias.

operational amplifier A high-gain feedback amplifier used as the basic unit of the analog computer and having its main application in function generation. The most usual configuration of the operational amplifier is illustrated. It can be shown that

$$E_o(s)/E_{in}(s) = -Z_f(s)/Z_i(s),$$

where $E_{in}(s)$, $E_o(s)$ are respectively the input and output voltages and $Z_i(s)$, $Z_f(s)$ are the input and feedback impedances, all of these being functions of the complex variable s. See ANALOG COMPUTER; SUMMER; INTEGRATOR; INVERTER.

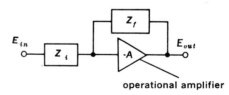

operational amplifier

If $A \gg 1$. $\dfrac{E_{out}}{E_{in}} \cong -\dfrac{Z_f}{Z_i}$

operational notation A method for transforming a DIF-FERENTIAL EQUATION into an algebraic equation by means of the operator s defined in such a way that

$$s = \frac{d}{dt}, \quad s^n = \frac{d^n}{dt^n}, \quad \frac{1}{s} = \int_0^t dt.$$

Working with a set of transformed equations is far simpler than working with the original set of differential equations. Applying the transformation to the voltage-current relationships of the three basic circuit elements, inductance (L), resistance (R), and capacitance (C), we get

$$v = sLi, \quad v = Ri, \quad v = \frac{1}{sC} i$$

Noting that $v = sLi$ closely resembles $v = j\omega Li$ (and likewise for the other elements), we call sL, R, and $1/sC$ the **operational impedances** for these elements. Similarly, we can define **operational admittance** and **operational immitance.** In the case where all initial conditions are zero the transformed equation is identical to the Laplace transform of the original differential equation where $V(s)$ and $I(s)$ are the Laplace transforms of $v(t)$ and $i(t)$ respectively. The time functions $i(t)$ and $v(t)$ are found by inverse transformation. For linear equations the assumption of zero initial conditions creates no hardship, as by the principle of SUPERPOSITION initial conditions can be represented by appropriate ideal generators. Care must be taken in commuting the operator s, as

$$sLi = L\, di/dt \neq i\, dL/dt = isL.$$

See LAPLACE TRANSFORM.

operation code In a digital computer, the part of an instruction that indicates what operation is to be performed.

operator *n.* In mathematics, a symbol that indicates the performance of one or more operations such as addition, multiplication, differentiation, integration.

optical axis 1 In a lens, a straight line joining the foci of the curved surfaces. 2 In a birefringent crystal, a line along which no birefringence occurs.

optical transistor A gallium arsenide device in which signals are carried in the form of light rather than as electrical energy.

optics *n.* The science of light and vision, concerned with the area of the electromagnetic spectrum with frequencies between those of x-rays and microwaves.

optoelectronic integrated circuit An INTEGRATED CIRCUIT combining optic and electric energy ports.

optophone *n.* A photoelectric device that can convert print into sound, enabling the blind to read.

orange peel A recording blank with a surface like that of an orange.

orbit *n.* The path described by a body around its center of attraction.—**orbital** *adj.*

order *n.* A way of describing or ranking things in terms of the number of times an operation is performed or is implied. For instance, a second-*order* differential equation is one that contains a second derivative, that is, the result of differentiating twice. A second-*order* physical system is one described by such an equation. An **order of magnitude** implies a tenfold difference between two quantities.

order of reflection See HOP.

orders of logic A measure of the speed with which a signal passes through a LOGIC CIRCUIT.

ordinary ray See DOUBLE REFRACTION.

ordinate *n.* The *y* coordinate of a point.

organ *n.* The logic section of a computer SUBASSEMBLY.

OR-gate A LOGIC CIRCUIT having two or more inputs and a single output. A signal appears at its output if any of its inputs are energized.

origin *n.* 1 In a system of coordinates, the point at which the axes intersect. 2 The ADDRESS at which a computer program or a block of data begins.

orthicon *n.* A storage television camera tube without the photo-cathode to image section of the IMAGE ORTHICON, having, therefore, less sensitivity.

orthiconoscope *n.* ORTHICON.

Orthonik *n.* A highly grain-oriented iron-nickel alloy. See PERMENORM 5000Z.

oscillate *v.* To repeat a cycle of motions or to pass through a cycle of states with, ideally, strict periodicity.

oscillation *n.* 1 The condition of that which oscillates. 2 A CYCLE.

oscillator *n.* Something that oscillates, in particular, a self-excited electronic circuit whose output voltage or current is a periodic function of time.—**oscillatory** *adj.*

oscillogram *n.* **1** A record made by an oscillograph. **2** A photograph of an oscilloscope display.

oscillograph *n.* A RECORDER that produces a plot of a varying current or voltage as a function of time.

oscilloscope *n.* An instrument in which the horizontal and vertical deflection of the electron beam of a cathode-ray tube are, respectively, proportional to a pair of applied voltages. In the most usual application of the instrument the vertical deflection is a signal voltage and the horizontal deflection is a linear time base.

osmium (Os) Element 76; Atomic wt. 190.2; Melting pt. 2700°C; Boiling pt. above 5300°C; Spec. grav. (20°C) 22.48; Hardness 7; Valence 2, 3, 4, or 8; Spec. heat 0.031 cal/(g)(°C); Electrical resistivity (0°) 9.5×10^{-6} ohm/cm. Used in making lamp filaments.

osophone *n.* A headphone transferring sounds directly to the bones of the head.

outphaser *n.* In electronic organs, a circuit that changes a sawtooth wave to something approaching a square wave by adding to the sawtooth a second sawtooth of twice the frequency and half the amplitude in reverse phase, thus cancelling the even harmonics. —**outphasing** *n.*

output *n.* **1** The useful energy or signal delivered by a network or device. **2** The terminals or port where the energy is delivered. **3** The information fed by a computer to its secondary storage or an external device.

output admittance A function Y relating the current and voltage at the output terminals of a network; the reciprocal of OUTPUT IMPEDANCE.

output impedance A function Z relating the current and voltage at the output terminals of a network, specifically,

$$v(t) = Zi(t).$$

output resonator A resonant cavity in a klystron which is excited by the velocity-modulated beam of electrons and transmits energy to an output waveguide.

overcoupling *n.* For two resonant systems, a degree of coupling that is more than CRITICAL COUPLING.

overcurrent *n.* An excessively large current.

overflow *n.* In a digital computer, a condition in which the number of digits that result from an operation exceeds the capacity of a REGISTER or a STORAGE LOCATION.

overhang *n.* The acid-resistant material remaining on a printed circuit after the etching process.

overload *n.* **1** A condition in which a system or device operates poorly or is damaged because one of its ratings, as current capacity, permissible input voltage, permissible output energy, etc., is being exceeded. **2** The influence or factor causing this condition.

overmoded *adj.* Having two or more MODES propagating in it at once, as a waveguide.

overmodulation *n.* A degree of *amplitude modulation* in which the amplitude of the modulating wave exceeds one-half that of the carrier.

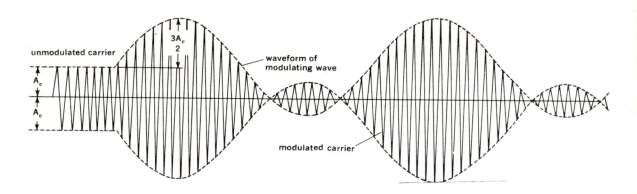

unmodulated carrier $\frac{3A_c}{2}$ waveform of modulating wave A_c A_c modulated carrier

override *v.*   To take manual charge of an automatic machine.

overshoot *n.*   A type of transient behavior in which the initial response to a change in input exceeds the steady-state response.

overtone *n.*   A tone in a complex sound having a higher frequency than the fundamental.

overvoltage *n.*   **1** The difference between electrode potential with and without electrolysis conditions. **2** In a Geiger-Muller counter, the amount of potential that exceeds the threshold voltage. **3** An excessive voltage.

O-wave   The ordinary component of a radio wave that has interacted with the ionosphere.

Owen bridge   An ac BRIDGE used to measure self-inductance in terms of capacitance and resistance.

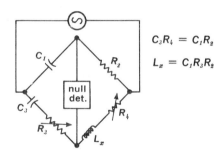

$$C_3 R_4 = C_1 R_2$$
$$L_x = C_1 R_3 R_2$$

oxidation *n.*   The process of combining with oxygen, more generally, the process by which atoms lose valence electrons or begin to share them with more electronegative atoms.

oxide *n.*   **1** Any binary compound of oxygen with another element. **2** In magnetic recording, the ferric oxide particles used on the magnetic tape.

oxygen (O)   Element 8; Atomic wt. 16.000; Melting pt. $-218.4°C$; Boiling pt. $-183.0°C$; Boiling pt. (1 atm pressure) $-182.97°C$; Spec. grav. ($-182.96°C$ liquid) 1.14; Density (0°) 1 429 g/l, (0°C, 1 atm, gas) 1.4920 g/l; Valence 2; Critical temp. $-118.8°C$; Critical pressure 49.7 atm; Triple pt. (solid, liquid, and gas in equilibrium) $-218.80°C$. Used for electrical welding processes.

ozone *n.*   An allotropic form of oxygen (O^3).

P

P *symbol*   **1** (P) Anode. **2** (*P*) Permeance. **3** (P) Phosphorus. **4** (p) pico- **5** (*P*) Power **6** (P) Primary winding.

PA *abbr.*   Pulse amplifier.

PABX *abbr.*   Private Automatic Branch Exchange.

pacemaker *n.*   Anything that acts to set the rate of the heartbeat, in particular, an electronic device producing a pulse output that stimulates the heart to contract, used in the treatment of various cardiac conditions.

packing density   In a digital information storage system, the number of bits stored per unit length or area of the storage medium.

pad *n.*   **1** An attenuator that has a fixed loss. **2** A resistive network connected between a source and load in order to match their respective output and input impedances. **3** One of the terminal connections of an encapsulated circuit.

padder *n.*   A small variable capacitor connected in parallel with the capacitor of a tuned circuit in order to allow fine adjustment of the resonant frequency.

pair production   A process in which a photon is transformed into a positron and an electron upon interacting with a strong electric field.

palladium (Pd)   Element 46; Atomic wt. 106.7; Melting pt. 1553°C; Boiling pt. 2200°C; Spec. grav. (20°C) 12.16; Hardness 4.8; Valence 2 or 4; Crystal structure Face-centered cubic; Spec. heat (0°) 0.0584 cal/(g) (°C); Thermal conductivity (20°C) 0.168 cal/(cm²)(°C) (sec); Coefficient of linear expansion (20°C) 12.4 $\times$ $10^{-6}/°C$; Modulus of elasticity 16 $\times$ 10^6 psi; Tensile strength (hard) 55,000 psi, (annealed) 25,000 psi; Electrical resistivity (20°C) 10.8 $\times$ 10^{-6} ohm/cm; Temp. coefficient of electrical resistivity (0–100°C) 0.0038/°C.

Palmer scan   In radar, a conical beam changing its azimuth constantly.

PAM *abbr.*   Pulse-amplitude modulation.

panoramic receiver   A radio-frequency spectrum analyzer consisting of a receiver whose tuning is periodically swept across a band of frequencies and whose output is displayed on an oscilloscope that has its time base synchronized with the receiver sweep.

paper tape Tape on which data can be recorded by means of punched holes.

parabola *n.* The locus of a point moving in a plane so that its distance from a fixed point and a fixed straight line are equal.

parallax *n.* The apparent displacement of an object when it is viewed successively from two points.

parallel 1 *n.* A connection of two or more circuit elements or branches so that they meet at common junction points and have the same voltage across them. **2** *adj.* Having to do with such a connection. **3** *adj.* Indicating a type of computer in which several operations are performed on the same or different data at once.

paralysis *n.* A condition in which the overloading of an amplifier causes its coupling capacitors to become charged, rendering it inoperative for a short but significant time after the overload.

paramagnetism *n.* A property of certain materials whereby they are magnetized parallel to a magnetic field in which they are placed and to an extent proportional to the strength of the field. The permeabilities of such materials exceed unity but, in general, do not approach those of ferromagnetic materials.—**paramagnetic** *adj.*

parameter *n.* **1** Any of the arbitrary constants used in writing an analytic expression. For instance in the expression

$$L\, di/dt + Ri = 0,$$

i is the independent variable while L and R are parameters. **2** A variable t such that given a function

$$y = f(x_1, x_2, \ldots x_n),$$

the function can be represented by the set of equations

$$y = g_0(t),\, x_1 = g_1(t),\, x_2 = g_2(t),\, \ldots x_n = g_n(t).$$

parametric amplifier A low-noise microwave amplifier in which energy supplied from an ac source (called the pump) at a frequency ω_p is coupled to an incoming signal of frequency ω_s. At least one modulated idler frequency ω_i is produced. Generally

$$\omega_i = \omega_p - \omega_s$$

but variations in which the device acts as a frequency converter are also possible. The hardware used ranges from electron beams modulated at the frequencies ω_s and ω_p, to nonlinear reactances that are made to vary at ω_p while the signal is passed through them.

parametron *n.* A resonant element similar to a parametric amplifier, used for information storage in certain digital computers. It consists of a resonant circuit in which one of the reactances can be varied according to an applied signal. If this signal is at twice the resonant frequency, oscillation at the resonant frequency is sustained by heterodyne action and is stable in either of two phases 180° apart. One phase represents a 0, the other a 1.

paramistor *n.* A digital logic device that contains several PARAMETRONS.

paraphrase amplifier An amplifier having two output ports, the signal at each port being a phase-inverted version of the signal at the other, often used to provide the excitation for push-pull stages.

parasitic *n.* An unwanted signal, oscillation, etc.

parasitic element An antenna element that is not directly connected to the feed line and whose action depends on electromagnetic coupling between it and a driven element.

partial *n.* In acoustics, a simple sinusoidal component of a complex tone.

partial derivative Of a FUNCTION of a set of independent variables $x_1, x_2, \ldots x_n$, the DERIVATIVE

$$fx_i(x_1, x_2, \cdots x_n) = Dx_i(x_1, x_2, \cdots x_n) = f_i(x_1, x_2, \cdots x_n)$$
$$= \partial f(x_1, x_2 \cdots x_n)/\partial x_i$$

taken with respect to one of these variables x_i, with all of the others held constant.

partial differential equation A differential equation that contains more than one independent variable and/or derivatives or differentials of more than one independent variable.

partial tone PARTIAL.

particle *n.* **1** A body having finite mass and negligible dimensions. **2** Any of the components of an atom.

partition noise In a pentode, a form of noise that results from the random division of current between the plate and screen.

Paschen's law   The statement that the sparking potential E_s of a gas is proportional to the product of the
pressure of the gas and the spacing of the electrodes
across which the potential exists; analytically

$$E_s = kpd,$$

where k is a constant of proportionality, p pressure,
and d electrode spacing.

passband *n.*   The range of frequencies that will pass
through a transducer with virtually no attenuation.

passivate *v.*   To treat the surface of (a semiconductor
chip) with a relatively inert material in order to protect
it from contamination.

passive substrate   A SUBSTRATE usually made of glass
or ceramic which supports but forms no part of the
circuit.

patch *n.*   **1** A temporary connection, as on a PATCH
BOARD. **2** In a computer, a section of coding inserted
into a routine to correct or change it. —**patch** *v.*

patch board   A panel having a number of jacks which
terminate circuits and which may be joined together in
different ways by short lengths of cable.

Pauli exclusion principle   The principle that no pair of
identical particles can simultaneously occupy the same
quantum state. The principle applies to electrons, protons, and neutrons and accounts for the shell config-
urations of extranuclear electrons and for the shell
structure of nuclei themselves.

P band   A band of microwave frequencies extending
from 225 Mhz to 390 Mhz.

PCM *abbr.*   **1** Pulse-code modulation. **2** Pulse-count
modulation.

P.D. *British abbr.*   Potential difference.

PDM *abbr.*   Pulse-duration modulation.

peak *n.*   A MAXIMUM or MINIMUM; an extreme value.

peak *v.*   **1** To increase or sharpen the peaks of (a waveform). **2** To broaden the frequency response of (an
amplifier) by including inductors in its coupling networks so as to cancel the input and output capacitances of its active elements.

peak factor   For a recurrent waveform, the ratio of the
peak value to the root-mean-square value.

peaking transformer   A transformer operated in such a

way that its core is saturated in one direction or the
other for most of a single ac cycle, with the result that
the secondary voltage waveform is sharply peaked at
each flux reversal. Sharpness of the peaking is enhanced by an approximately rectangular hysteresis
loop in the core.

peak-overshoot-ratio   In a feedback control system, a
performance criterion defined as the ratio of the error
at the first OVERSHOOT to the initial magnitude of the
error to which the overshoot is a response.

peak point   In a characteristic curve of an active device,
a relative maximum or an extreme value.

pedestal *n.*   **1** A flat-topped pulse elevating the base
level for another signal. **2** The level of the synchronizing pulses in a composite television picture signal.

Peltier effect   The change in temperature produced at
the junction of two dissimilar metals upon the passage
of an electric current, the direction of the change depending on the direction of the current.

PEM *abbr.*   Photoelectromagnetic effect.

pencil tube   A long, thin vacuum tube used as a uhf
oscillator or amplifier.

penetration depth   **1** In induction heating, the effective
depth of the induced current. The SKIN EFFECT causes
this to be nearer the surface with high frequencies than
with low frequencies. **2** The extent to which an external
magnetic field penetrates a SUPERCONDUCTOR.

penetration factor   A vacuum-tube parameter, defined
as the reciprocal of the AMPLIFICATION FACTOR.

penta- *combining form*   Five. Also, before a vowel
pent-; *pentode.*

pentagrid converter   A seven element vacuum tube
whose cathode and first two grids act as an oscillator,
while the other three grids act as a mixer. The enclosure of these two elements of a superheterodyne receiver in one envelope involves a significant saving.

pentatron *n.*   A five-element vacuum tube combining
two triodes with a shared cathode and providing push-
pull amplification within a single envelope.

PEP *abbr.*   Peak effective power.

perceptron *n.*   An adaptive system that can, at least in
theory, classify a set of patterns according to their
salient common features, thus enabling itself to rec-

ognize patterns that despite a number of nonsignificant variations in form are essentially equivalent.

period *n.* The time that elapses between any two successive similar phases of an oscillation or other regular, cyclical motion.

periodic *adj.* Recurring at definite intervals.

Permalloy *n.* An alloy of iron (20 to 60%) and nickel (40 to 80%), used as a soft magnetic material. **45 Permalloy** (45%Ni) has an initial permeability (μ_0) 2,500 gauss/oersted, a maximum permeability (μ_M) 25,000, a coercive force of 0.1 oersted, and a Curie point (θ) of 400°C. **78 Permalloy** (78.5%Ni) has μ_0 of 8,000 gauss/oersted, μ_M of 100,000, a coercive force of 0.05 oersted, and θ at 600°C. **4–79 Permalloy** (17%Fe, 4%Mo, 79%Ni) has μ_0 of 20,000 gauss/oersted, μ_M of 100,000, a coercive force of 0.05 oersted, and θ at 460°C. 4 Mo Permalloy has the same characteristics as HYMN 88. See SUPERMALLOY.

permanent magnet A magnet that retains its magnetism indefinitely after being magnetized.

permanent-magnet moving coil instrument A device having a coil pivoted on bearings between the poles of a permanent magnet. The coil is restrained by springs which also serve as leads and is deflected, when a current is passed through it, to an angular position proportional to the current. When the coil is put in series with a high resistance the amount of current passed is minimal and is proportional to the applied voltage.

permatron *n.* A two-element, gas-discharge tube in which an external magnetic field controls the onset of conduction.

permeability *n.* A parameter μ of a material, equal to the ratio of the magnitude of the magnetic induction **B** in the material to the magnitude of the magnetic field strength **H**. Analytically,

$$\mu = |\mathbf{B}|/|\mathbf{H}|.$$

A distinct but related parameter is **relative permeability** μ_r, given by

$$\mu_r = \mu/\mu_0,$$

where μ_0 is the permeability of free space.

permeance *n.* In a portion of a magnetic circuit bounded by a pair of equipotential surfaces and by another surface tangent at every point to the direction of magnetic induction, the ratio of the flux ϕ through a cross section to the difference of magnetic potential (MAGNETOMOTIVE FORCE) $\mathcal{F}$ between the bounding surfaces; analytically permeance P is given by $P = \phi/\mathcal{F}$.

Permendur *n.* A soft magnetic alloy of 50% cobalt and 50% iron having high permeability at high flux density. Its initial and maximum permeabilities are 800 and 5,000 gauss/oersted, its coercive force 2 oersteds and it has the saturation induction of 24,500 gauss. Its Curie point is 980°C.

Permenorm 5000Z An alloy of 50% nickel and 50% iron with an initial permeability of 4,000 gauss/oersted and a maximum permeability of 70,000; coercive force 0.05 oersteds, saturation induction 16,000 gauss and Curie point 500°C.

Perminvar *n.* A soft magnetic alloy of 30% iron, 45% nickel, and 25% cobalt having high and constant permeability.

permittivity *n.* The degree to which a given dielectric can store electrostatic energy. The ratio of the capacitance of a capacitor with the given dielectric to the capacitance of a similar capacitor with a vacuum dielectric; dielectric constant.

persistence *n.* Of the phosphor of a cathode ray tube, the relation between its radiant emissive power and the time elapsed after excitation has ceased.

persistor *n.* A cryogenic, thin-film storage device consisting of a superconductive inductor connected in parallel with a normally superconductive switching device which becomes resistive when the current reaches a certain level. A persistent current is established in the loop consisting of the inductor and resistor by a 'write' pulse. Readout is often accomplished by quenching this current and detecting the voltage transient across the inductor. See CRYOTRON.

persistron *n.* A solid-state display panel that amplifies light by the use of electroluminescent and photoconductive devices.

perturbation theory A set of methods for the approximate analysis of the behavior of a system under small changes in its independent variables, useful when the equations describing the system prove intractable.

perv *n.* The measure of PERVEANCE, defined as equal

to the perveance of an electron tube in which the plate current is $v^{3/2}$ when the effective plate voltage is v.

perveance n.   A proportionality constant G relating the plate and cathode voltage e_p of a vacuum tube to the space charge limited plate current i_p. The value of G depends upon the geometry of the system. For a triode the plate current is given by

$$i_p = G \left(e_g + \frac{e_p}{\mu} \right)^{3/2}$$

where e_g is the grid to cathode voltage and μ the amplification factor. For a diode (where there is no grid)

$$i_p = G e_p^{3/2}.$$

pf, PF abbr.   1 Picofarad. 2 Power factor.

PFM abbr.   Pulse frequency modulation.

PG abbr.   Pulse generator.

phanotron n.   A gas-filled rectifier tube having a hot cathode.

phantastron n.   A circuit having a pentode connected so as to act both as a MONOSTABLE MULTIVIBRATOR and a MILLER INTEGRATOR. In response to a trigger a rectangular pulse is generated internally, appearing, due to the integrator action, as a RAMP at the plate.

phantom channel   A communications channel for which no independent conductive path exists. The signal information for such a channel is added to that of other channels in such a way that no additional paths are required but all signals are recoverable with negligible interaction.

phase n.   1 Of a simple harmonic motion of the form

$$u = \cos(\omega t + k\phi)$$

the angle $\omega t + k\phi$. The phase of a sinusoidal function of time relates the function to a fixed instant of time or to another sinusoid of the same frequency. It is usual to express phase as an angle θ such that $0 < \theta < 2\pi$. Thus $\theta = \omega t + k\phi - n2\pi$ where n is such an integer as is needed to place θ in the desired interval. 2 Any of the single circuits comprising a polyphase ac system.

phase angle   In an electric circuit, an angle ϕ that indicates the amount by which the current lags or leads the voltage, given by

$$\phi = \tan^{-1} (\text{Im } i(t)/\text{Re } i(t)),$$

where Im $i(t)$ and Re $i(t)$ are respectively the imaginary and real components of the current. See COMPLEX NUMBER; IMPEDANCE.

phase constant   The imaginary part of the PROPAGATION CONSTANT.

phase control   A method of controlling the amount of ac power delivered to a load by allowing conduction to take place for just a fraction of each cycle. The control elements used are THYRATRONS and THYRISTORS.

phase distortion   The alteration of a complex waveform produced as it passes through a network or transducer whose phase shift is a function of frequency.

phase inverter   1 A network or device such as a paraphrase amplifier, which produces two output signals that differ in phase by half a cycle. 2 A network that changes the phase of a wave by half a cycle.

phase modulation   A form of ANGLE MODULATION in which the phase of the carrier is varied according to the information to be transmitted. Analytically, if a modulation function $m(t)$ is defined, a phase-modulated carrier $P(t)$ is given by

$$P(t) = A_o \cos [\omega_o t + \Delta \theta \, m(t)],$$

where $\Delta \theta$ defines the amount of phase shift per unit change in the modulation function. As frequency is the derivative with respect to time of phase, phase modulation and frequency modulation are closely related.

phase reversal   A 180° change in phase such as a wave might undergo upon reflection under certain conditions.

phase-shift discriminator   A circuit that produces an output proportional to the phase difference between two input signals. When used as an fm demodulator the input is a set of push-pull signals and a reference voltage 90° displaced from each of them. These are taken across tuned circuits in such a way that the phase difference between the push-pull signals and the reference is very nearly proportional to the difference between the input frequency and the resonant frequency of the tuned circuits.

phase-shift oscillator   An oscillator consisting of a phase-inverting amplifier whose output is coupled to its input through a network that produces a phase change that is an odd multiple of 180°. Generally this change is a function of frequency; thus if the amplifier gain makes up the network losses there is a single frequency at which the BARKHAUSEN CRITERION is satisfied.

phase velocity The velocity with which a point where there exists an electromagnetic wave of a certain fixed phase, moves through space in the direction of propagation of the wave.

phasitron *n.* An electron tube designed to frequency-modulate an externally generated rf carrier. Internal electrodes are arranged to produce a corrugated disk of electrons that rotates inside the tube at the carrier frequency. The modulation is applied to an external coil that produces an axial magnetic field. This field, by producing angular displacement of the electron disk, frequency- (and phase-) modulates the carrier.

phasor *n.* **1** A graphical representation of a sinusoidal function of time as a directed line segment plotted on the complex plane. Thus since the exponential

$$Ae^{j\phi} = A(\cos \phi + j \sin \phi)$$

it is convenient to plot it in this form. **2** In radio broadcasting, a network used to shift the phase of the excitation of an antenna element to the value needed for a desired directional response.

phenolic *n.* A thermosetting plastic derived from the reaction of phenol with an aldehyde, used extensively in radio, telephone, and other electronic equipment.

phi (written Φ, ϕ) The twenty-first letter of the Greek alphabet, used symbolically to represent any of various coefficients, constants, etc., of which in electronics the principal ones are: **1** (ϕ) Angle. **2** (ϕ) Electron affinity. **3** (ϕ) Function of. **4** (Φ) Magnetic or radiant flux. **5** (ϕ) Phi polarization.

Phillips gauge A vacuum gauge that measures gas pressure as a function of the current in a glow discharge.

Phillip's screw A screw with a cruciform recess.

phon *n.* The unit of loudness level. The loudness level of a sound in phons is equal to the sound pressure level in decibels re 0.0002 microbar of a pure tone of 1000 hz which a group of listeners judge to be equally loud. See SONE.

phone HEADPHONE.

-phone *combining form* Voice; sound: used in names of musical instruments and other sound-transmitting devices.

phono PHONOGRAPH.

phonograph *n.* A record player.

phono jack A jack designed to accept a phono plug.

phonon *n.* A conceptual quantum of acoustic energy whose introduction into the theory of crystal lattice vibrations allows great formal simplification. The energy of such a quantum is $h\nu$, where h is Planck's constant and ν is the frequency of the sound wave.

phono plug A type of small plug commonly used on shielded audio cables.

phosphor *n.* Any luminescent material.

phosphor bronze An alloy of copper, tin, and phosphorus (95.5 Cu, 4 Sn, 0.5 P is a common percentage) used for contact springs in switches.

phosphorescence *n.* Luminescence which persists more than 10^{-8} second after excitation has ceased.

phosphorus (P) Element 15; Atomic wt. 30.975; Melting pt. 44.1°C; Boiling pt. 280°C; Spec. grav. (yellow) 1.82, (red) 2.20; Valence 3 or 5; Spec. heat 0.177 cal/(g)(°C); Electrical conductivity (0–20°C) 10^{-11} mho/cm. Used as a donor impurity in semiconductors.

phot *n.* The cgs measure of luminance, defined as one lumen/cm².

photo- *combining form* Light; of, pertaining to, or produced by light.

photocathode *n.* An electrode which emits electrons when exposed to light.

photocell *n.* A PHOTOELECTRIC CELL.

photoconductor *n.* A substance whose conductivity increases under the influence of light.

photoelectric *adj.* Of or pertaining to electrical or electronic effects due to the action of light.

photoelectric cell **1** A solid-state electron device whose current-voltage characteristics are affected by light. **2** An electron tube containing a photocathode that emits electrons when exposed to light or radiation of similar wavelengths.

photoemissive *adj.* Emitting electrons on exposure to light or radiant energy of a frequency close to that of light.

photon *n.* A quantum of electromagnetic energy equal to $h\nu$, where h is Planck's constant and γ the frequency of the radiation.

photopositive *adj.* Of a material, increasing in conductivity under the action of light.

photosensitive *adj.* Having physical characteristics that vary under the influence of light.

phototransistor *n.* A semiconductor device in which the current resulting from hole-electron pairs generated by radiant energy acts like the base current in an ordinary transistor.

phototube *n.* An electron tube containing a PHOTOCATHODE.

photovaristor *n.* A VARISTOR in which the current-voltage characteristics are modified by the action of light.

photovoltaic *n.* Generating an electromotive force as a result of exposure to light.

pi (written Π, π) The sixteenth letter of the Greek alphabet, used symbolically to represent any of various coefficients, constants, etc., of which in electronics the principal ones are: **1** (π) the ratio of the circumference of a circle to its diameter; 3.14159 . . . **2** (Π) Peltier coefficient. **3** (Π) Hertzian vector. **4** (π) Poynting vector. **5** (Π) Product of.

pickup *n.* **1** A transducer that converts sonic, visual, or other data into electrical signals. **2** The minimum power that will trip a relay or cause a silicon-controlled rectifier to conduct.

pico- *combining form* One trillionth (10^{-12}) of a specified quantity or dimension.

picofarad *n.* One millionth of a microfarad.

Pierce oscillator A COLPITTS OSCILLATOR in which a piezoelectric crystal is placed in the feedback loop between the anode and the grid.

piezo- *combining form* Pressure.

piezoelectricity *n.* A property of certain dielectric crystals whereby a difference of electric potential is developed across them as a result of applied mechical stresses. The effect is reciprocal in that the application of an electric field to such a crystal causes it to develop mechanical strains. **—piezoelectric** *adj.*

piezoid *n.* A piezoelectric crystal prepared for use in a circuit.

piezo-junction transistor A transistor having a mechanical linkage applied to its emitter-base junction in such a way as to cause the conductivity to vary with pressure, the variations of current appearing in amplified form at the collector.

pigtail *n.* A short, flexible, braided wire used to connect a stationary and a moving part.

pile-up **1** In a pulse amplifier, a shift in the base line as a result of pulses arriving so closely spaced that coupling capacitors have insufficient time to discharge fully, or that the amplifier cannot recover from nonlinear operation. **2** A STACK.

pillbox antenna A CHEESE ANTENNA.

pinch effect The constrictive force exerted on a conductor as a result of the magnetic attraction between parallel components of the current it carries. It follows from AMPERE'S LAW that the pressure P at the surface of a cylindrical wire is given by

$$P = I^2/200\pi r^2 \text{ dynes/cm}^2,$$

where I is the current and r the radius of the wire. While negligible at small currents, the effect increases with the square of the current and is further compounded by deformation of the conductor. Attempts have been made to use the pinch effect to confine plasmas in thermonuclear reactors.

pinchoff *n.* In a FIELD-EFFECT TRANSISTOR, a condition in which the gate bias causes the depletion region to extend completely across the channel, with a resulting cessation of drain current.

pi network Essentially, a variation of a DELTA NETWORK in which two junction points are respectively input and output terminals and the third junction point is a common.

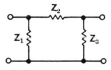

ping *n.* A sonic or ultrasonic pulse emitted by SONAR equipment.

ping-pong-ball effect   The repeated switching of sound between one loudspeaker and another in a stereophonic system.

pin jack   A small single-conductor JACK.

pin junction   A variation of a PN-JUNCTION in which the p- and n-type semi-conductor regions are separated by an intrinsic region.

pip n.   The representation of an echo on a radarscope.

pitch n.   The subjective attribute of a sound that undergoes the most marked variation in response to changes in frequency.

planar network   A NETWORK that can be drawn on a plane in such a way that no two BRANCHES cross.

planar transistor   A junction transistor manufactured by a process in which the surface of a chip is passivated with a thin film of oxide, dopants being introduced by successive etching and diffusion.

planchet n.   A metal container for radioactive materials.

Planck radiation formula   A formula that gives the spectral distribution of the radiation from a black body, namely the relation

$$E_\lambda d\lambda = \frac{hc^3}{\lambda^5} \cdot d\lambda^{-1} \left[\exp\left(\frac{hc}{k\lambda T}\right) - 1 \right]$$

where E_λ is the radiation intensity between the wavelength limits λ, $\lambda + d\lambda$, h is PLANCK'S CONSTANT, c is the speed of light, k is BOLTZMANN'S CONSTANT, and T is absolute temperature.

Planck's constant   A universal constant (symbol h) having the value of approximately 6.624×10^{-34} joule-second.

plane polarization   See POLARIZATION.

planetary electron   An electron that is in orbit round a nucleus. See FREE ELECTRON.

plan position indicator   A horizontal, radial-scan radar in which the area surrounding the set and any targets in it are displayed as on a map.

plasma n.   A gas that, while strongly ionized, contains positively and negatively charged particles in approximately equal numbers and is therefore electrically neutral.

plasma frequency   The natural oscillation frequency ω_ρ of a PLASMA composed of equal and oppositely charged types of particles. In radians per second, this frequency is given by

$$\omega_\rho = \sqrt{\frac{\rho q}{\varepsilon m^*}},$$

where ρ is the charge density of each type of particle q/m^* is the sum of the charge to mass ratios of the various particles, and ε is the relative permittivity (dielectric constant) of the surroundings.

plasmatron n.   A hot-cathode gas-discharge tube in which conduction between anode and cathode is through an independently generated PLASMA. Modulation of the anode current is accomplished by varying either the conductivity of the plasma or its effective cross section.

plate n.   In an electron tube, the main anode.

platinotron n.   A microwave amplifier tube requiring the presence of an external magnetic field for its operation. It is somewhat similar to a magnetron, but has no resonant circuit and has two rf ports rather than one.

platinum (Pt)   Element 78; Atomic wt. 195.23; Melting pt. 1773.5°C; Boiling pt. 4300°C; Spec. grav. (20°C) 21.37; Hardness 4.3; Valence 2; Spec. heat 0.032 cal/(g)(°C); Electrical resistivity (0°C) 10.6×10^{-6} ohm/cm. Used in contact points in switches and relays, for electric-furnace windings, in resistance thermometers, and grids of special purpose vacuum tubes.

playback n.   The realization of a recording as actual sound.

playthrough n.   An audible output that appears despite the fact that the volume control is set to its zero position.

plug n.   A device on the end of an electric cord that will fit into a JACK.

plugboard   PATCHBOARD.

plug-in   A module that can be plugged into a larger system by means of pins, sockets, or other appropriate connectors.

plumbing n.   Microwave WAVEGUIDES resembling pipes.

Plutonium (Pu)   Element 94; Atomic wt. 242; Valence

3, 4, 5, or 6; Radioactive element with a half-life of about 24,300 years; emits alpha particles. Used as a nuclear fuel.

PM *abbr.* **1** PHASE MODULATION. **2** PERMANENT MAGNET.

pn junction An interface at which there is a sharp transition between a region of P-TYPE SEMICONDUCTOR and N-TYPE SEMICONDUCTOR, forming essentially a diode. During normal operation a good approximation to the voltage-current characteristics is given by

$$I(V) = I_0(e^{\alpha V} - 1),$$

where α and I_0 (called the saturation current) are nominally constants, thus accounting for the nonlinear conduction and the familiar rectifying property. In practice α and I_0 may depend on temperature and applied voltage. For values of reverse bias greater than the Zener breakdown voltage (V_z) the exponential voltage-current relationship no longer applies and the diode develops an approximately constant voltage, thus acting as a voltage regulator. For excessive forward voltages the pn diode acts more like a resistor than a diode. The presence of a DEPLETION REGION between the n- and p-type regions results in a junction capacitance which varies inversely with the width of the depletion region and thus depends, not linearly, on the applied voltage. See DIODE, TRIODE, ZENER VOLTAGE.

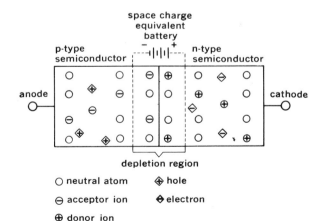

space charge
equivalent
battery

p-type
semiconductor   n-type
semiconductor

anode   cathode

depletion region

○ neutral atom   ⊕ hole

⊖ acceptor ion   ◈ electron

⊕ donor ion

Simplified diagram of a pn junction under zero bias.

pnpn device A semiconductor device containing three pn junctions, equivalent essentially to a pnp transistor and an npn transistor arranged in a regenerative feedback configuration. The device conducts when sufficient current is applied to the bases of the transistors to make the product of their current gains (β) greater than 1. Some pnpn devices can be turned off by reducing one of these base currents sufficiently; others require interruption of the supply voltage. See SILICON CONTROLLED RECTIFIER; SILICON CONTROLLED SWITCH; GATE-CONTROLLED SWITCH.

pnp transistor A bipolar transistor in which the emitter and collector are of p-type semiconductor material and the base is n-type.

point effect The crowding together of the lines of force in the electric field from the surface of a charged conductor in a region where the surface comes to a point or has a sharp curvature. This is a result of the fact that the lines of force must be normal to the surface.

point source A source of radiation that is sufficiently small with reference to the observer to be considered as a point.

point-to-point radio communication Communication between two fixed stations.

poison *n.* An impurity that impairs the operation of a system or device, as a nonfissionable neutron absorber in a nuclear fuel, an impurity that reduces the electron emission of a thermionic cathode, etc.

polarity *n.* The characteristic of having an axis with reference to which certain physical properties are determined; charge, for instance, may be positive or negative and thus has polarity.

polarization *n.* **1** A process by which the charge carriers (ELECTRONS, IONS, HOLES, etc.) of opposite signs are made to migrate toward different regions of a body under the influence of an external field. **2** A vector **P** representing the electric dipole moment per unit volume of a dielectric. Analytically **P** is given by

$$\mathbf{P} = (1/\gamma)(\mathbf{D} - \varepsilon_0\mathbf{E}) = 1/\gamma \cdot \varepsilon_0(\kappa' - 1)\mathbf{E},$$

where **D** is electric displacement, **E** electric field strength, ε_0 the permittivity of space, κ' the dielectric constant of the material, and γ a geometrical constant of proportionality. **3** A condition in which the ELECTRIC FIELD VECTOR **E** of an electromagnetic wave is predictable in magnitude and direction rather than random. In **linear polarization** the direction of **E** is fixed while its magnitude varies sinusoidally at the frequency of the wave. In **circular polarization** the magnitude of **E** is constant while its direction rotates through a full circle during each period of the wave. **Elliptical polarization** is a more general case of circular polarization in which the magnitude of **E** is greater in, say, the x direction than in the y direction. **Plane polarization** is a case in which all of the waves comprising a single wavefront have their electric field vectors in the same plane. **4** In an electric cell, the formation of products around the electrodes in such a way as to degrade the performance of the cell.

polarize v.   Undergo or cause to undergo POLARIZATION.

polar modulation   A form of amplitude modulation in which the positive excursions of the carrier are modulated by one signal and the negative ones by another.

pole n.   **1** A point at which the denominator of a function becomes zero; more precisely, a point z_0 at which a function of the COMPLEX VARIABLE $z = x + iy$ can be represented in the form

$$f(z) = g(z)/(z - z_0)^k$$

where k is a positive integer called the order of the pole, $g(z_0) \neq 0$, and $g(z)$ is continuous and has a derivative at z_0 and in the neighborhood of z_0. When $k = 1$, z_0 is referred to as the simple pole of $f(z)$. The name derives from the fact that a three-dimensional plot of $f(z)$ would appear to 'climb a pole' of infinite length at z_0. **2** A MAGNETIC POLE. **3** Either terminal of an electric cell.

pole face   The face of a magnet nearest the attracted object.

polonium (Po)   Element 84; Atomic wt. appr. 210. Radioactive element; emits alpha particles; used in static eliminators.

poly-   combining form Many; several.

polycrystal n.   A crystal structure which is nonhomogeneous, but which is composed, rather, of very small crystals with definite boundaries separating them.

polyethylene n.   A strong flexible thermoplastic with very high dielectric strength, a low dielectric constant and a very low dissipation factor, used on transmission lines.

polyphase adj.   Denoting an ac system that consists essentially of a number of interconnected circuits whose supply voltages are displaced in phase. Normally, for a system of n phases the displacement is $\phi = 2\pi/n$. Most commonly n has the value 3.

polyplexer n.   In radar, a DUPLEXER that also has lobe switching capabilities.

polyrod n.   A dielectric antenna made of polystyrene rods.

polystyrene n.   A glasslike thermoplastic material which has high insulation resistance, good dielectric strength, and high refractive index. It is one of the stiffest of plastics.

pool cathode   Liquid mercury used as a cathode in a mercury arc tube. The pool of mercury serves both as the cathode and the supplier of vaporized mercury for the arc. It is continually replenished by condensation.

port n.   A means by which energy may be coupled into or out of a system, often, in the case of an electrical system, consisting of a pair of terminals.

portable standard   A readily transportable physical standard used for calibration of instruments in the field.

positive adj.   **1** Greater than zero. **2** Having a deficiency of electrons.

positive transmission   In television, a system in which the amplitude of the video signal is directly proportional to brightness.

positron   See ELECTRON.

pot n.   POTENTIOMETER.

pot v.   To encapsulate in a container filled with insulating material.—**potting** n.

potassium (K; Lat. Kalium.)   Element 19; Atomic wt. 39.10; Melting pt. 62.3°C; Boiling pt. 760°C; Spec. grav. (20°C) 0.87; Density (100°C) 0.819 g/cm², (400°C) 0.747 g/cm², (700°C) 0.676 g/cm²; Hardness 0.5; Valence 1; Heat of fusion (63.7°C) 14.6 cal/g; Heat of vaporization (63.7°C) 496 cal/g; Heat capacity (200°C) 0.19 cal/(g)(°C); Viscosity (70°C) 5.15 millipoises; Thermal conductivity (200°C) 0.017 cal/(sec)(cm²)(cm)(°C); Vapor pressure (342°C) 1 mm, (696°C) 400 mm; Surface tension (100–150°C) About 80 dynes/cm; Electrical resistivity (150°C) 18.7 × 10⁻⁶ ohm/cm. Used on cathodes of photo-tubes.

potential *n.* The work per unit of interaction with the field done in moving a body from a reference point, generally infinity, to a point of interest (*P*) in a field of conservative force. If $\mathbf{A(s)}$ is a vector representing the strength of the field at a point, the potential *V* is given by

$$V = \int_\infty^P \mathbf{A(s)} \cdot \mathbf{ds}.$$

where **ds** is a vector element of path length from ∞ to *P*. See ELECTRIC POTENTIAL; MAGNETIC POTENTIAL.

potential barrier **1** A region of space in which there is a relative MAXIMUM of potential which acts (with certain exceptions) as a barrier to particles whose energies do not exceed its magnitude. See TUNNEL EFFECT. **2** A DEPLETION REGION.

potential energy The energy stored in a system by virtue of its configuration, more precisely, in a system of conservative forces initially in configuration *a* and finally in configuration *b*, potential energy $K_{a,b}$ is given by

$$K_{a,b} = -W_{a,b},$$

where $W_{a,b}$ is the work done in moving the system from one configuration to the other. It can be shown that this result is independent of the paths followed by any constituents of the system.

potentiometer *n.* **1** A resistor provided with a tap that can be moved along it in such a way as to put the tap effectively at the junction of two resistors whose sum is the total resistance, the ratio of the two effective resistors being a function of the position of the tap. **2** A measuring instrument in which a potentiometer (def. 1) is used as a voltage divider in order to provide

a known voltage that can be balanced against an unknown voltage.—**potentiometric** *adj.*

pothead *n.* A piece of equipment through which cables are brought from an underground cable into the open.

Potier diagram A graph describing the capabilities of an ac generator, comparing variations in field current with load and voltage.

power *n.* **1** The rate, with respect to time, at which WORK is done, that is, power *P* is given by $P = dW/dt$, where *W* is work and *t* time. Electrically, power is generally measured in watts. Instantaneously, it is given by the product of current and voltage. **2** The value of an EXPONENT.

power amplifier The final stage in a multistage amplifier circuit, designed to give power to the load, rather than mainly as a voltage amplifier.

power factor The cosine of the phase angle *θ* in an ac circuit. It can be shown that the mean power *P* in such a circuit is given by

$$P = EI\cos\theta,$$

where *E* and *I* are, respectively, root-mean-square values of voltage and current.

power series A series of the form

$$\sum_{n=0}^{p} a_n x^n,$$

where *p* is finite or infinite. Series of this kind are useful for representing or evaluating certain functions.

power transistor A transistor designed to handle large currents and safely dissipate large amounts of power.

Poynting's vector A vector **P** whose outward component normal to a closed surface in an electromagnetic field represents, when integrated over that surface, the net outward energy per unit time through the surface. Analytically

$$\mathbf{P} = \mathbf{E} \times \mathbf{H},$$

where **E** is electric field strength and **H** magnetic field strength. For the transmission of energy along a coaxial cable it can be shown that

$$|\mathbf{P}| = I^2R$$

and the direction is along the axis of the cable.

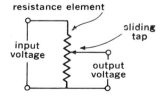

P–P, PP *abbr.* Peak to peak.

PPI *abbr.* Plan position indicator.

PPM *abbr.* Pulse-position modulator.

PPPI *abbr.* Precision plan position indicator.

pps *abbr.* Pulse(s) per second.

praseodymium (Pr) Element 59; Atomic wt. 140.92; Melting pt. 940°C; Spec. grav. (20°C) 6.5; Valence 3, 4, or 5; Spec. heat 0.048 cal/(g)(°C); Electrical conductivity (0–20°C) 0.015 × 10⁶ mho/cm.

preamp PREAMPLIFIER.

preamplifier *n.* An AMPLIFIER primarily intended to operate with a low-level input signal, increasing the signal to supply an input for other amplifier circuits.

preburning *n.* The stabilization of an electron tube by heating, with or without current flow.

precision *n.* The exactness of a number, determined by the number of significant digits.

preconduction current In a thyratron or other grid-controlled gas tube, the small plate current that flows before the grid is brought to firing voltage.

precursor *n.* An UNDERSHOOT.

preemphasis *n.* The process of increasing the magnitude of certain frequency components in a signal with respect to the others; in particular, the increase in certain frequencies, usually the higher, employed to improve the signal-to-noise ratio in fm·radio transmission. A complementary de-emphasis network at the receiver can restore the signal to its original form.

preferred numbers A series of values conventionally adopted as standards for capacitors and resistors to maintain an orderly progression of sizes. The American Standards Agency has adopted $\sqrt[5]{10}$ and $\sqrt[10]{10}$ as bases for the series; the Radio Manufacturers Association, $\sqrt[6]{10}$, $\sqrt[12]{10}$, and $\sqrt[24]{10}$. The latter series is used chiefly for small radio components.

preselection *n.* In digital computers, a technique whereby data from the next input tape is stored while the computer is still processing other data.

preselector *n.* A tuned preamplifier placed between an antenna and the input terminals of a radio receiver in order to improve the selectivity and sensitivity of the receiver.

presence *n.* In a sound reproduction system, the illusion of the performer's presence, achieved in high fidelity systems by a subtle exaggeration of the response of the system around 5,000 hz.

pressure *n.* **1** A STRESS whose salient property is that it is the same in all directions. Its dimensions are, like those of other stresses, force per unit area. **2** Sound pressure.

pressure microphone A nondirectional microphone in which the electrical response is approximately proportional to the sound pressure of the wave.

pressure pad A felt pad used in some tape recorders to press the magnetic tape against the recording or reproducing head.

pretransmit-receive tube An auxiliary transmit receive tube used to protect the main one against excessive power and to protect the receiver against spurious frequencies.

PRF *abbr.* Pulse repetition frequency.

primary *n.* A PRIMARY WINDING.

primary cell A cell in which the energy of one or more chemical reactions is converted to electrical energy, and which, despite the theoretical reversibility of the chemical reactions, is not designed to be efficiently recharged.

primary colors Any three colors that when mixed in suitable proportions produce any color. Primary colors may be subtractive, where the primaries absorb colors from white light (e.g. magenta, cyan, yellow; red, blue, yellow) used in developing color photography; or additive, where the primaries form a color by the addition of their light (e.g. red, green, blue).

primary service area The area within which the GROUND WAVE of a broadcast station can be received without objectionable interference or variation in strength.

primary standard A physical standard that is established and maintained by some authority and which is used for calibrating secondary standards.

primary winding In a multiwinding transformer, the winding that is connected to the input terminals.

primitive *n.* A FUNCTION $F(x)$ whose DERIVATIVE is a function $f(x)$ under consideration, that is, if $F'(x) = f(x)$, then $F(x)$ is the primitive of $f(x)$.

primitive period For a periodic quantity $y = f(x)$ the smallest value of k such that $f(x) = f(x + k)$.

principal axis 1 An axis chosen as a reference line for a coordinate system used in plotting the directional response or characteristics of a system or transducer. 2 The OPTICAL AXIS of a crystal. 3 The longest axis of a crystal.

printed circuit An insulated board on which connections, and sometimes resistors, have been deposited by any of several processes.

print-through The transfer of signal between adjacent layers of a reel of magnetic tape, resulting, on playback, in pre- and post-echos of heavily recorded passages.

probe *n.* In an electrical measuring instrument, a rod-like part that is held in the hand and positioned so that a conductor at one end is coupled to the circuit under measurement. In some applications the probe itself contains a network that performs part of the processing of the measured signal.

product demodulation The demodulation of a product-modulated signal by passing it through a device whose output is the product of a locally generated signal of carrier frequency and the sidebands produced by the original PRODUCT MODULATION. The output contains the original modulation together with other components that are easily filtered out.—**product demodulator.**

product modulation A form of amplitude modulation in which the output is equal to the product of the carrier and the modulating signal; analytically, with a modulation function defined as $m(t)$, the modulated carrier $P(t)$ is given by

$$P(t) = m(t)A_0 \cos \omega t.$$

It is characteristic of this technique that the carrier is suppressed when there is no modulation. The spectrum consists of the sidebands alone.

profile chart A vertical cross section of the terrain between a microwave transmitter and receiver, used in determining antenna height requirements.

program 1 *n.* A sequence of operations performed by a system in order to accomplish some desired result, in particular, the sequence of operations or instructions executed by a digital computer in solving a prob-

lem. 2 *v.* To prepare a program for a computer or other system.—**programmer** *n.*

program relay A multipole relay that when energized produces a preset switching sequence, typically by means of a cam-operated mechanism.

promethium (Pm) Element 61; Atomic wt. approx. 145; Valence 3; All known isotopes are radioactive.

promethium cell A nuclear energy cell in which beta particles from promethium 147 strike a phosphor and cause it to glow, the light output being converted to electricity by photocells.

propagate *v.* To pass, as a wave, through a medium or along a transmission line.—**propagation** *n.*

propagation anomaly A discontinuous change in the propagation characteristics of a medium.

propagation constant A characteristic γ of a transmission line or medium that indicates its effect on a wave that is propagating through it. For example the voltage (or current) x received a distance d from the sending end of a transmission line is given by

$$x = x_0 e^{-d\gamma},$$

where x_0 is the wave amplitude at the sending end and γ is a complex number given by

$$\gamma = \alpha + j\beta.$$

The attenuation in nepers per unit distance is given by α; the phase shift per unit distance is given by β. In a transmission line it can be shown that

$$\gamma = \sqrt{ZY}$$

where Z is series impedance per unit distance and Y shunt admittance per unit distance. Similarly in a vacuum

$$\gamma = \sqrt{\mu_0 / \varepsilon_0},$$

where μ_0, ε_0 are respectively the permeability and permittivity of a vacuum. The term propagation *constant* is somewhat of a misnomer, as γ, in general, can vary with position, direction, and frequency.

propagation ratio The ratio of the complex magnitude of the ELECTRIC FIELD VECTOR of a wave at one point to that at a second point.

protactinium (Pa) Element 91; Atomic wt. 231. All

known isotopes are radioactive. Pa-231 has a half-life of about 34,000 years; emits alpha particles.

proton *n.* A positively charged particle having a rest mass of 1.6724×10^{-24}g, a charge of magnitude equal to the electron (1.60×10^{-19} coulomb) and a spin of one-half unit, that is of $\hbar/2$ or $h/4\pi$, where h is PLANCK'S CONSTANT.

proximity effect The change in the current distribution within a conductor as a result of the field that surrounds another current-carrying conductor nearby.

PS *abbr.* Power supply.

pseudovector *n.* A quantity whose properties are the same as those of a vector except that it changes sign if the handedness of the coordinate system is reversed. The **vector product** of two vectors is a typical pseudovector.

psi (*written* Ψ, ψ) The twenty-third letter of the Greek alphabet, used symbolically to represent any of various coefficients, constants, etc., of which in electronics the principal ones are: **1** (ψ) Function. **2** (ψ) Time dependent wave function. **3** (Ψ) Total electric flux. **4** (Ψ) Displacement flux.

psopho- *combining form* Noise.

psophometric *adj.* Of or having to do with the measurement of noise.

Pt-Co An alloy of 77% platinum and 23% cobalt with the highest coercive force of any alloy, 4,000 oersteds.

PTM *abbr.* Pulse-time modulation.

p-type semiconductor A semiconductor material in which due to the presence of ACCEPTOR IMPURITIES the MAJORITY CARRIERS are positive HOLES.

pull *v.* **1** To cause (an oscillator) to depart from its designed frequency of operation. **2** To depart from a designed frequency of operation, as an oscillator.

pull-in current The minimum amount of current needed to trigger a relay.

pulsating direct current A direct current whose magnitude varies with time.

pulse 1 *n.* A wave form consisting of the sum of a positive step function and a negative step function separated in time. **2** *v.* To supply with energy in pulses. —**pulser** *n.* See bottom of page for diagram.

pulse mode A finite, coded sequence of pulses, used to select and isolate a communication channel.

pulse moder A device used to produce a PULSE MODE.

pulse modulation A type of modulation in which the

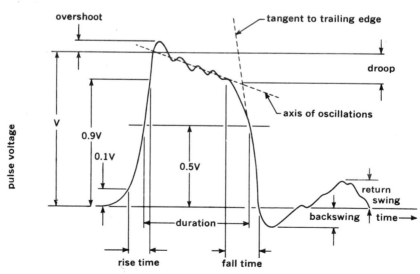

information signal is sampled at appropriate intervals and its instantaneous amplitude used to control some characteristic of a pulsed transmission of the carrier.

pulse packet   The volume of space in which the energy of a radar pulse is contained.

pulse stretcher   A circuit that when triggered by a pulse of a given amplitude and duration produces an output pulse whose amplitude is proportional to that of the original but whose duration is longer.

pump *v.*   To supply high-frequency energy to (a laser, maser, parametric amplifier, etc.).

punch-through   A form of transistor failure in which an internal short develops between emitter and collector across the base, usually as a result of excessive voltages.

puncture 1 *n.*   The breakdown of a dielectric under an electrical stress. **2** *v.* To cause or undergo a puncture.

Pupin coil   Any of a set of inductors introduced into a transmission line at regular intervals in order to adjust its CHARACTERISTIC IMPEDANCE and PROPAGATION CONSTANT.

pure tone   A sound wave whose pressure (VOLUME VELOCITY) is a simple sinusoidal function of time.

purity *n.*   In color television, the degree to which a primary color contains no admixture of the other primaries.

push-pull *adj.*   **1** Indicating a BALANCED network, particularly one that contains active components. **2** Indicating the balanced excitation signal supplied to an active push-pull network.

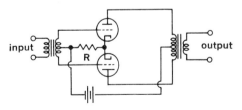

A triode push-pull amplifier

push-push circuit   An amplifier circuit containing a pair of similar active elements connected so that their

inputs are fed in phase opposition and their outputs are delivered in parallel to a common load, usually used as a frequency multiplier.

pW, pw *abbr.*   Picowatt(s).

PWM *abbr.*   Pulse-width modulation.

Q

Q *symbol.*   **1** (Q) QUALITY FACTOR. **2** (Q,q) Electric charge. **3** (Q) Thermoelectric power. **4** (Q) QUADRATURE.

QCW signal   The QUADRATURE-PHASE SUBCARRIER SIGNAL.

Q-multiplier *n.*   A device that multiplies the QUALITY FACTOR, and thus the selectivity, of a circuit, used often in the i-f section of radio receivers.

Q signal   Any of a set of sequences of three letters beginning with Q used as standard abbreviated messages as given in the International List of Abbreviations for Telegraphy.

QSL card   A card sent by one ham to another confirming that they have actually been in radio contact.

quad *n.*   **1** In a cable, a structural unit consisting of four conductors twisted together, each being separately insulated. **2** A series-parallel or parallel series combination of four elements, as semiconductors, passive components, or even complete circuits. **3** A form of antenna consisting of a driven element and a reflector each a square .25 λ long on a side with a separation of .15–.20 λ between the two, where λ is the wavelength at the operating frequency.

quadrant *n.*   A quarter section of a circle; an arc of 90°.

quadrature *n.*   A condition in which a pair of alternating quantities differ in phase by 90°.

quadrature-phase subcarrier signal   In color television, the QUADRATURE component of the 3.579 Mhz chrominance subcarrier.

quality factor   A figure of merit Q_0 for a resonant circuit or system. In particular, a system in which $Q_0 \gg 1$ exhibits sharp frequency selectivity and tuning characteristics. In terms of circuit elements

$$Q_0 = \omega_0 L/R$$

for a series *RLC* circuit, and

$$Q_0 = \omega_0 RC$$

for a parallel *RLC* circuit, where ω_0 is the resonant frequency of the circuit. In terms of network frequency characteristics

$$Q_0 = \omega_0/B$$

where *B* is the half power bandwidth of the network, that is

$$B = \omega_2 - \omega_1$$

where $\omega_{2,1}$ are the upper and lower CORNER FREQUENCIES of the system.

quantize *v.* To operate on a continuous function in such a way as to convert it into a discrete set of values connected by step functions, that is, for example, given a continuous function $f(x)$ and a second function $g(x)$ such that if

$$a_i \leq f(x) < a_{i+1}, \; g(x) = a_i,$$

where a_i is one of a set of n values differing by Δa, and it is understood that $f(x)$ is sucessively evaluated at $x_0, x_0 + \Delta x, \ldots x_0 + n\Delta x, \; g(x)$ is a quantized version of $f(x)$. Clearly $g(x)$ is an approximation of $f(x)$, the closeness of the approximation depending on the magnitudes of Δa, Δx. Thus the transformation of $f(x)$ into $g(x)$ may irreversibly destroy some of the information in $f(x)$. See SAMPLING THEOREM.—**quantization** *n.*—**quantizer** *n.*

quantized *adj.* Restricted to a discrete set of values.

quantum *n.* An indivisible packet of energy. Given a wave of frequency ν, the magnitude of its energy quantum is $h\nu$, where *h* is Planck's constant; the quantum of angular momentum is

$$\hbar = h/2\pi.$$

quantum efficiency The efficiency with which energy quanta are used in producing some desired result, defined, in general, as the ratio of the number of events of the desired type to the number of quanta absorbed, with the implicit assumption that the magnitude of each quantum is known.

quantum number Any of a set of numbers assigned to the particular values of a quantized quantity in its discrete range. The state of a particle or system may be described by a set of compatible quantum numbers.

quantum state The state of a particle or system as described by its quantum numbers.

quantum theory A physical theory originally formulated to explain certain effects that were anomalous in the light of classical physics. Central to this theory is the notion that certain physical quantities are restricted to discrete sets of values, that is, that they are quantized. Another of its features is the duality that it attributes to particles, treating them as waves in some cases and, indeed, as particles in others. Being concerned mainly with the microscopic properties of matter, quantum theory is rarely used in the design of electronic circuits but rather in the development of electronic devices, microcircuits and thin film devices being examples of such developments. See QUANTUM; BAND THEORY OF SOLIDS; TUNNEL EFFECT.

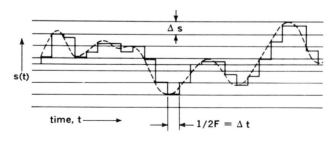

A quantized waveform

quantum transition   A transition between two quantum states.

quarter-wave *n.*   One quarter cycle of a wave.

quartz *n.*   A naturally occurring form of silicon dioxide SiO_2 that forms in hexagonal, positive, uniaxial crystals. It shows a fairly strong piezoelectric effect and rotates the plane of polarization of light clockwise or counterclockwise depending on the particular crystal.

quasi-optical *adj.*   Of waves, behaving in a manner similar to light, as, for instance, being restricted by the horizon in their propagation.

quench *v.*   To terminate (a discharge, oscillation, etc.).

quiescent *adj.*   Of an amplifying device, in a state in which no input signal· is being received.

quieting sensitivity   In an fm tuner or receiver, the minimum input signal for which the improvement output signal-to-noise ratio is greater than or equal to a specified figure.

R

R *symbol*   **1** (*R, r*) RESISTANCE. **2** (*r*) Radius.

racon *n.*   A radar beacon that receives and retransmits radar signals for navigation purposes.

rad *n.*   A unit of nuclear radiation equivalent to 100 ergs of absorbed energy per gram of absorbing material.

rad *abbr.*   RADIAN(S).

radar *n.*   A system of radio detection and ranging which detects objects by beaming rf pulses that are reflected back by the object and measures its distance by the time elapsed between transmission and reception. The strength of the echo signal is determined by the **radar equation**

$$W_R = W_T \frac{G^2 \lambda^2 \sigma}{(4\pi)^0 R^4}$$

where W_R is received echo power. W_T is transmitted power, G is antenna gain, λ is radar carrier wavelength, σ is target cross section, and R is range.

radar beacon   See RACON.

radar clutter   Unwanted echoes which interfere with observation of desired signals on radar indicators.

radar fence   A line of radar stations designed to give warning of enemy air attack.

radar nautical mile   The time that a radar signal takes to hit a target one nautical mile away and return; 12.261 microseconds.

radar paint   A paint that absorbs and does not reflect radar radiation.

radarscope *n.*   The oscilloscope of a radar receiver.

radiac *n.*   A device for the detection and measurement of nuclear radiation.

radial lead   A lead coming out of the side of a component.

Resistor with radical leads

radian *n.*   A unit of angular measure, defined as equal to the angle that intercepts on the circumference of a circle an arc equal in length to the radius. A radian is considered the ratio of two lengths and is therefore dimensionless. 1 radian = 57° 17′44.80625″.

radiance *n.*   The radiant intensity of a surface in a given direction; equivalent to luminance but including all forms of radiation.

radiant energy   Energy transmitted in the form of waves, especially electromagnetic waves.

radiant flux   The time rate of flow of radiant energy expressed in various units.

radiate *v.*   To emit rays.

radiation *n.*   The emission and propagation of RADIANT ENERGY.

radiation pattern   A graphical representation of the reception of an antenna.

radiation resistance   The ratio of the total power radiated by an antenna to the effective antenna current as measured at the feed point.

radio *n.* The propagation, transmission, and detection of RADIO WAVES.

radioacoustic position finding A method of discovering the distance from a fixed point by comparing the difference in time between the reception of echoes from an explosion and a radio signal.

radioacoustics *n.* The science of transmission of sound by radio.

radioastronomy *n.* A branch of astronomy in which celestial objects are studied by their radio emissions rather than by their optical emissions.

radio-frequency *n.* Any wave frequency lying between 10 khz and 30,000 Mhz.

radio-frequency choke A small air- or ferrite-core inductor used to present a high impedance to radio-frequency currents.

radiopaque *adj.* Impenetrable by x-rays or other radiation.

radiophare *n.* A radio station sending out a constant signal enabling the pilot of an aircraft to fix his bearing towards it.

radio relay system A broadcasting system in which intermediate stations are used between the studio and ultimate transmitter.

radiosonde *n.* A lightweight miniature weather station carried aloft, generally by balloon, and radioing meteorological information to the ground.

radiothermics *n.* The use of radio waves to generate heat as in DIATHERMY.

radiotransparent *adj.* Transparent to radiation; invisible in x-ray photography or fluoroscopy.

radio wave An electromagnetic wave whose frequency is in the range from 10^3 to about 3×10^{11} hertz.

radium (Ra) Element 88; Atomic wt. 226.05; Melting pt. 960°C; Boiling pt. 1140°C; Spec. grav. 5 (?); Valence 2; Ionic radius (Ra^{++}) 2.45Å (estimated); magnetic susceptibility Feebly paramagnetic. Radioactive element with a half-life of 1622 years; emits alpha particles, yields radon.

radix *n.* A number or symbol used as the base of a scale of enumeration.

radix complement See COMPLEMENT.

radix-minus-one complement See COMPLEMENT.

radome *n.* A housing for a radar antenna, made of a dielectric material that passes rf-energy with negligible or small attenuation.

radon (Rn) Element 86; Atomic wt. 222; Melting pt. −100°C; Boiling pt. −61.8°C; Spec. grav. 9.73 g/l; Valence 0. Radioactive gaseous element with a half-life of 3.825 days; emits alpha particles, yields radium A.

radux *n.* A hyperbolic navigation system employing continuous wave phase comparison.

railing *n.* The jamming of enemy radars by transmissions whose pulse repetition rate is 50–150 khz, causing patterns that resemble fence railings to appear on the display.

rainbow generator A device which can generate signals to reproduce the whole color spectrum on a color television picture tube.

ramark *n.* A fixed, continuously-operating radar beacon for ships and aircraft.

ramp *n.* A function that changes value at a constant rate; a straight line of finite slope; a SWEEP.

Ramsauer effect The abnormally low attenuation of a low-energy electron beam as it passes through any of the inert gases.

random access In a computer, direct access to information in the storage without the necessity of sequential consideration of other storage items.

random variable A variable having a set of possible values X_i such that each member of the set occurs with a definite probability.

range *n.* **1** The limits between which variation is possible. **2** The set of all possible values taken by a DEPENDENT VARIABLE.

raster *n.* Scanning pattern of a television picture tube.

rate action Action that is proportional to the rate at which a variable changes, that is to say, proportional to its derivative with respect to time. The concept is generally applied to control systems, where the variable of interest is an error signal.

rate effect A break into conduction by a THYRISTOR when forward voltage is applied too rapidly between its anode and cathode.

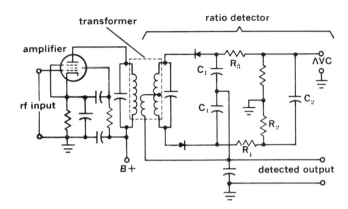

ratio detector   An fm detector consisting essentially of a PHASE-SHIFT DISCRIMINATOR that has been modified to provide limiting as well. Its name derives from the fact that the sum of the outputs of the diodes is practically constant over its usable range, the recovered modulation being derived from variation in the ratio of the diode voltages.

ray *n.*   A line of propagation of any form of radiant energy.

raydist *n.*   A hyperbolic navigation system employing three stations to give an accurate fix on a vessel.

Rayleigh disc   An acoustic radiometer consisting of a light disc suspended by one of its edges, used in measuring particle velocity.

Rayleigh scattering   The scattering of light as it passes through a medium composed of fine particles. For particles that are small compared with λ the wavelength of the light, the light scattered in a direction that makes an angle θ with the direction of incidence is proportional to $(1 + \cos^2\theta) \lambda^{-4}$.

RBE *abbr.*   Relative biological effectiveness.

RC, R–C *abbr.*   **1** Resistor-capacitor. **2** Resistance-capacitance. **3** Resistance-coupled.

RCTL *abbr.*   Resistance-capacitance transistor logic.

R-display   A type of A-display in which an echo can be expanded to allow closer observation.

reach-through voltage   In a transistor, that voltage which, when applied from collector to base, causes the space charge layer to expand into the emitter junction.

reactance *n.*   **1** The imaginary part of IMPEDANCE for a network in the sinusoidal steady state. It follows from the equations describing the voltage-current relationship of an ideal inductor (L) that the current it passes in the sinusoidal steady state is given by $I = E/j\omega L$. Thus, as impedance is the ratio of voltage to current, it follows that the impedance of an ideal inductor is given by

$$Z = j\omega L.$$

Since Z is here purely imaginary it is called **inductive reactance** and given the symbol X_L. Similarly, it is possible to derive for an ideal capacitor a **capacitive reactance**

$$X_0 = 1_c/j\omega C$$

2 A CAPACITANCE or INDUCTANCE. **3** ACOUSTIC REACTANCE.—**reactive** *adj.*

reactance modulator   A frequency modulator consisting essentially of a radio-frequency oscillator whose tuned circuit contains a reactance tube (transistor) whose gain is varied according to the modulating signal, thus shifting the frequency of the oscillator.

reactance tube (transistor)   A vacuum tube or transistor connected so as to appear as a reactance to the rest of the circuit. This is accomplished by deriving the grid (base) excitation from the plate (collector) voltage through an RC network that changes its phase by approximately 90°. Thus, since the plate (collector) current is in phase with the grid (base) excitation the tube (transistor) draws an apparently reactive current. The magnitude of the apparent reactance is controlled by variation of the gain of the tube (transistor).

reactive power See COMPLEX POWER.

reactor *n.* A circuit element designed primarily to exhibit a certain reactance; a capacitor or inductor.

read around number The number of times an item in the storage section of a computer can be consulted without affecting neighboring areas.

readout *n.* In a computer, that part of the output which is temporary and may be directly interpreted. Internally, the computer makes its own readout of auxiliary storage devices.

readthrough *n.* The recording on magnetic tape and subsequent analysis of enemy radar signals in order to determine their bearing, pulse width and other characteristics.

real *adj.* Denoting a number whose square is positive, that is a number a such that $0 < a^2$. See COMPLEX NUMBER.

real time operation The use of a computer to control a process as it is actually occurring, necessitating, in general, relatively rapid operation on the part of the computer.

recalescent point The temperature at which a ferromagnetic material in the process of cooling will temporarily cease the process and even gain heat. The phenomenon is due to changes in the crystal structure which give off heat.

receiver *n.* A device for the reception and, if necessary, demodulation of electronic signals.

reception *n.* The quality of reception of any form of emission in a particular device or area.

reciprocal **1** *n.* Of a number a, a second number a^{-1} such that $a \cdot a^{-1} = 1$. **2** *adj.* Exhibiting reciprocity, as a system, network, transducer, etc.

reciprocal impedance Either of a pair of impedances $Z_{1,2}$ that are related to another impedance Z_0 in such a way that

$$Z_1 Z_2 = Z_0^2.$$

reciprocity *n.* A property of a physical system whereby interchange of its input and output does not affect the system response to a given excitation. Consider in particular the four terminal network illustrated. The network exhibits reciprocity if

$$v_1/i_2 = z_{12} = z_{21} = v_2/i_1,$$

provided that z_{12} is measured with the network connected as shown and z_{21} is measured with the load and excitation interchanged.

reciprocity theorem for electric networks A theorem stating that in an electric network composed exclusively of passive, bilateral, linear impedances the ratio of a voltage introduced into one branch to the current produced in another branch is unchanged if the positions of the current and voltage are interchanged.

recombination *n.* A process in which current carriers of opposite signs combine and form stable, neutral entities.

recombination coefficient A coefficient that gives the rate of recombination of ions in a gas. The coefficient depends on the nature of the gas and its pressure.

recombine *v.* To undergo RECOMBINATION.

record *n.* **1** A grooved disc that reproduces sounds on a phonograph. **2** In a computer, a unit of information made up of smaller units.

recorder *n.* A device that puts electrical signals into permanent form, involving, for example, a magnetic tape or wire, a recording disk, a film, or an OSCILLOGRAPH.

recording *n.* The process of registering a relatively permanent physical record of sounds or other communicable signals.

recording head **1** A MAGNETIC HEAD. **2** A cutting stylus for records.

record player A motor-driven turntable with a pickup attachment and auxiliary equipment for the playing of phonograph records.

recovery *n.* In an electronic device, the time required to enable the device to react to new signals.

rectifier *n.* A device which has the ability to pass current in one direction and not the other.

rectify *v.* To change an ac current into dc.—**rectification** *n.*

redundancy *n.* **1** In electronic devices, the employment of several tubes, transistors, etc. to perform the same function, thus increasing reliability. **2** In information theory, the extent to which a communications source overspecifies information. More precisely, redundancy *r* is given by

$$r = 1 - H/H_{max},$$

where H is the information rate of the channel and H_{max} is the theoretical maximum information rate of the channel when coded in the same way. Broadly speaking, redundancy makes it increasingly likely that errors in reception will be detected.

reed relay or **reed switch** A switching device consisting of magnetic contactors sealed into a glass tube, the contactors being actuated by the magnetic field of an external solenoid, electromagnet, or permanent magnet.

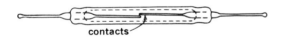

contacts

reflect *v.* To turn (energy) back.

reflectance *n.* Of a surface, the ratio of the reflected radiant flux to the incident radiant flux.

reflected impedance The impedance seen at the input of a network when its output is terminated in an impedance of a specified value. Most often, the 'network' referred to in this connection is a transformer.

reflection *n.* The turning back of all or part of radiant or other energy when it encounters a discontinuity in the medium in which it is traveling.

reflection coefficient For electromagnetic energy traveling through a medium or along a transmission line, the ratio of the amplitude reflected wave to that of the incident wave when the incident wave comes upon a discontinuity in the medium or line. If the measured characteristic of the wave is voltage, or its analog, ELECTRIC FIELD STRENGTH, the reflection coefficient is given by

$$A_r/A_0 = (Z_2 - Z_1)/(Z_2 + Z_1),$$

where $A_{0,r}$ are respectively the amplitudes of incident and reflected voltages (electric field strengths) and $Z_{1,2}$ are respectively the characteristic impedances on the incident and emergent sides of the discontinuity. See CHARACTERISTIC IMPEDANCE, CHARACTERISTIC WAVE IMPEDANCE.

reflection factor REFLECTION COEFFICIENT.

reflectometer *n.* A device for measuring the incident and reflected waves in a transmission line and, often, the time lapse between them.—**reflectometry** *n.*

reflector *n.* In an antenna, a PARASITIC ELEMENT located in a direction other than that of the major lobe.

reflex circuit A single stage of amplification that operates on two signals in widely separated frequency ranges.

reflex klystron A KLYSTRON in which one cavity acts as both buncher and catcher.

refraction *n.* The change of direction of a ray in passing from one medium to another of different density. —**refract** *v.*; **refractive** *adj.*

refractive index The ratio of the velocity of a specific radiation in a vacuum to its velocity in a given medium.

refractometer *n.* An instrument for measuring refractive indices.

regenerate *v.* To restore (a deteriorated pulse) to its original waveshape.

regeneration *n.* POSITIVE FEEDBACK.

regenerative feedback Positive FEEDBACK.

register *n.* In a computer, a small storage device for temporary use during arithmetic operations.

regulate *v.* To stabilize the voltage of a power supply. —**regulation** *n.*

regulator *n.* A device for keeping at constant strength the current (voltage) produced by a generator, alternator, or other source.

rejector *n.* TRAP.

rejector circuit A circuit that rejects signals of a certain frequency.

rejuvenator *n.* A device for restoring the emissivity of a thermionic cathode by running it at an elevated temperature for a short period of time.

rel *n.* A unit of RELUCTANCE, defined as one AMPERE-TURN per MAXWELL.

relative address In a computer routine, the address of a particular item relative to the origin of the routine.

relativistic particle A particle whose velocity is of a magnitude sufficient to cause its MASS to exceed its rest mass by a significant amount.

relativity *n.* A principle that states the impossibility of determining the absolute motion of a physical body or frame of reference through space, and as a result of which matter, space, and time must be considered interdependent. In particular, the **general theory of relativity,** dealing with coordinate frames in arbitrary states of relative motion, postulates that (1) the laws of physics have the same mathematical form in all coordinate systems, and (2) that gravitational fields and inertial fields cannot be experimentally distinguished. It follows from these postulates that space-time is curved. The **special theory of relativity,** dealing with coordinate systems that are in relative motion with uniform velocity along a straight line, postulates that (1) the description of a physical phenomenon can in no way depend on the absolute translational velocity of the coordinate frame, and (2) that the speed of propagation of an electromagnetic wave is constant with respect to all coordinate frames that are in translational motion along a straight line with constant speed, and does not depend on the speed of the source. As a result of these postulates time, mass, and dimension in the direction of motion are different in the two coordinate frames. An important consequence of the theory pertaining to electronics arises in the case of electrons that are accelerated to high velocities. —**relativistic** *adj.*

relaxation circuit A circuit in which a timing interval is established by the charging of an RC or LC time constant, the interval being terminated when the stored energy is more or less abruptly discharged. A **relaxation oscillator** is an oscillator circuit whose period is controlled by such a time constant. In general, relaxation oscillators produce complex output waveforms.

relaxation time The time required for a system to approach equilibrium after a STEP-FUNCTION change in an externally applied perturbation. More precisely, assuming that the response of the system is exponential, the time required for the system to reach $1 - 1/e$ of its final equilibrium after perturbation has been applied, or the time required for the response to decay to $1/e$ of its initial value after perturbation has ceased.

relay *n.* An electrically operated mechanical switch.

reliability *n.* The probability that a component or a device will perform satisfactorily.

relocate *v.* In a digital computer, to modify (the addresses within a routine) in such a way as to move it to another area of storage, thus allowing its use in conjunction with another routine that would otherwise occupy the same storage area.—**relocatable** *adj.*

reluctance *n.* A measure of the opposition that a MAGNETIC CIRCUIT presents to magnetic flux, defined as the reciprocal of PERMEANCE.

reluctivity *n.* The specific reluctance of a material, defined as the reciprocal of its magnetic PERMEABILITY.

remanence *n.* RESIDUAL INDUCTION.

remote cutoff tube A vacuum tube in which, due to the fact that the spacing of the wires in the control grid is not uniform, the AMPLIFICATION FACTOR μ depends on the choice of operating point. The closely spaced wires exert strong control over the electrons in the tube, thus giving a high value of μ. Increased negative bias drives this section of the grid into cutoff, effectively transferring control to the wider-spaced and, therefore lower-μ section.

repeater *n.* A device for automatically retransmitting electromagnetic signals.

repeating coil An audiofrequency transformer used in telephony.

resistivity *n.* Specific resistance, the reciprocal of CONDUCTIVITY.

resistor *n.* A circuit element designed to have a resistance of a given magnitude.

resnatron *n.* A CAVITY RESONATOR tetrode capable of

generating large power at high frequency and great efficiency.

resolution *n.* **1** The degree of detail in a spatial pattern, as on the face of a cathode ray tube. **2** The degree to which significant signals can be extracted from comparatively random signals, as with a radio telescope.

residual flux density RESIDUAL INDUCTION.

residual induction The MAGNETIC INDUCTION that remains in a sample of magnetic material after a saturating magnetizing force is removed.

resistance *n.* **1** The property of a circuit element by which it radiates energy away from a circuit when a current is passed through it. Analytically, resistance is given by

$$R = E^2/P,$$

where E is the impressed voltage and P is the power removed from the circuit. Since $P = EI$, it follows at once that $R = E/I$, and, further, that resistance is the real part of IMPEDANCE. See OHM'S LAW. **2** An ideal RESISTOR. **3** ACOUSTIC RESISTANCE.—**resistive** *adj.*

resistance box A device containing several resistors which can, singly and in combination, be connected into circuits for test purposes.

resolver *n.* A function generator, often of an electromechanical variety, whose output is proportional to the sine or cosine of its input.

resonance *n.* A condition in which an oscillatory system is subjected to an excitation that corresponds to one or more of its natural MODES. Electrically, the simplest case arises in a series (parallel) circuit of inductance (L), capacitance (C), and resistance (R). In a circuit of this kind, for some value of the frequency (ω_e) of a sinusoidal excitation, one of the IMMITANCES reaches a maximum while its reciprocal reaches a minimum. If R becomes negligibly small, the value of the minimum approaches zero. Referring specifically to the impedance of the series circuit

$$Z = R + j[\omega_e L - (1/\omega_e C)],$$

it is apparent that for some value of ω_e that the imaginary term vanishes and $Z = R$. Thus, at resonance, as R approaches zero, Z approaches zero, while admittance (Y) approaches infinity. It can be shown further that the sharpness of the maximum (minimum) de-

pends also on the resistance, or, more precisely, on the quality factor Q_0, given, in the case of the series circuit, by

$$Q_0 = \omega_0 L/R,$$

where ω_0 is the natural frequency of the circuit. In particular, if we define $k = \omega_e/\omega_0$,

$$Z/R = 1 + jQ_0 (k - 1/k).$$

For certain applications the transient behavior of resonant systems is of interest. This in general is determined from the transient part of the solution of the differential equation

$$Ly'' + Ry' + (1/C)y = f(t),$$

where $f(t)$ is the excitation. For any periodic excitation

$$f(t) = \sum_{n=1}^{\infty} A_n \sin(n\omega_e t + \psi_n).$$

It can be shown that the amplitude of the transient depends on the magnitude of each of the A_n and on the angles ϕ_n. The decay time of the transient depends on the amount of DAMPING present. Distributed systems such as recurrent networks and transmission lines also exhibit free modes. The frequency of each mode is given by

$$\omega_0 = nv/2l,$$

where v is the speed with which a wave travels through the system, l is the length of the system, and n is a positive integer. Through various means, any or all of these modes may be damped. In general, the transient behavior of such a system is extremely complicated. While nonelectrical systems also exhibit resonance, it can be shown through DYNAMIC ANALOGIES that the resonant behavior of most systems is the same.

resonant cavity CAVITY.

resonant circuit A circuit exhibiting RESONANCE or ANTIRESONANCE.

resonant gate transistor A field-effect transistor having a minute, resonant strip of gold coupled to its gate. In microcircuitry, this device makes its possible to achieve a tuned response without bulky inductors.

response *n.* A quantitative expression of the output of a device as a function of the input. See TRANSFER FUNCTION; TRANSFER IMPEDANCE.

responsor *n.* A receiver used with a TRANSPONDER.

rest mass The MASS that a particle appears to have when at rest with respect to the observer.

resultant *n.* An entity or quantity that is the result of some process or set of interactions.

retentivity *n.* In a magnetic material, a property that is measured by the RESIDUAL INDUCTION.

retrace *n.* FLYBACK.

retrofit *v.* To fit or be compatible with an earlier system or technology.

reverberate *v.* **1** To undergo or cause to undergo prolongation as a result of repeated reflections (too closely spaced to be perceived as individual echoes) within the boundaries of an enclosed volume, as a sound. **2** To treat (an audio signal) so as to synthetically create prolongation of this kind.—**reverberation** *n.*

reverberation time Time required for sound in a room to diminish by 60 dB after the source has cut off.

reverberator *n.* Any of various electromechanical devices designed to process an audio signal in such a way as to simulate the effects of reverberation.

reverse bias A voltage applied to a diode or pn junction with such polarity that conduction is negligible.

reversible counter A COUNTER that can count either forward or backward.

RF *abbr.* Radiofrequency.

RFC *abbr.* Radiofrequency choke.

RFI *abbr.* Radio-frequency interference.

rhenium (Re) Element 75; Atomic wt. 186.31; Melting pt. 3000°C; Boiling pt. 5900°C; Spec. grav. (20°C) 20.53; Valence 4, 5, 6, 7, 8; Spec. heat 0.033 cal/(g)(°C); Electrical conductivity (0–20°C) 0.051×10^6 mho/cm.

rheo- *combining form* Current or flow.

rheostat *n.* A resistor whose value can be changed by means of a mechanical control.

rho (written P, ρ) The seventeenth letter of the Greek alphabet, used symbolically to represent any of various constants, coefficients, etc., of which in electronics the principal ones are: **1** (ρ) Radius of curvature. **2** (ρ) Reflectance. **3** (ρ) Reflectivity. **4** (ρ) Resis-tivity or specific resistance. **5** (ρ) Volume electric charge density.

rhodium (Rh) Element 45; Atomic wt. 102.91; Melting pt. 1985°C; Boiling pt. above 2500°C; Spec. grav. (20°C) 12.5; Density (20°C) 12.41 g/cm³; Crystal structure Face-centered cubic, a = 3.80 at 20°C; Spec. heat (0°) 0.059 cal/(g)(°C); Thermal conductivity (20°C) cal/(cm)(cm²)(°C)(sec); Coefficient of linear thermal expansion (20°C) 8.3×10^{-6}°C; Modulus of elasticity 40×10^6 psi; Tensile strength (hard) 365,-000 psi, (annealed) 138,000 psi; Electrical resistivity (20°C) 4.5×10^{-6} ohm/cm. Used as cladding on sliding electrical contacts.

rhombic antenna A highly directional long-wire antenna in which the conductors are arranged in the form of a rhombus, the feed being generally applied at one of the acute angles. The radiation pattern is bidirectional, becoming unidirectional if the end remote from the feed is terminated in a resistor equal to the impedance of the antenna.

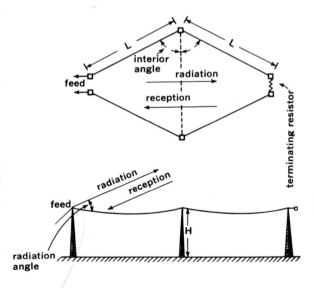

rhumbatron *n.* A CAVITY resonator.

RIAA curve A curve adopted by the Recording Industry Association of America as defining the standard recording characteristics for long-play records, the inverse of this curve being used for equalization on playback.

ribbon microphone A type of VELOCITY MICROPHONE in which a flat metallic strip suspended in a magnetic field acts as a diaphragm and as a single-turn moving coil.

RID *abbr.* Radio Intelligence Division (of the Federal Communications Commission).

ride gain To monitor the volume level of an audio-frequency circuit making adjustments as necessary to keep the volume level in its optimum range.

Rieke diagram A LOAD DIAGRAM presented in POLAR COORDINATES.

rig *n.* An amateur radio station.

right hand rule FLEMING'S RULE.

ring *v.* To undergo RINGING.

ring counter A counting circuit consisting of *N* bistable outputs arranged so that one and only one output is 'on' at a particular time. In response to an input pulse the 'on' condition is shifted to the next stage in the sequence, the first stage being the successor of the last.

ringing *n.* An oscillatory transient response occurring in an underdamped circuit or system as a result of a sudden change in input.

ring modulator Any of various balanced modulator or demodulator circuits in which the central operating element is a four-diode bridge circuit.

RIOMETER *n.* A meter for measuring the relative opacity of the ionosphere by measuring the incidence of cosmic noise in the HF and VHF regions.

ripple *n.* A residual ac component in the output of a dc power supply.

ripple voltage The alternating component of the unidirectional voltage from a rectifier or generator.

rise time The time required for a system response to a step input to go from 10 to 90 percent of its final value.

rising sun magnetron A multicavity MAGNETRON having resonators for two different frequencies arranged alternately for MODE separation.

rising sun resonator The anode of a RISING SUN MAGNETRON.

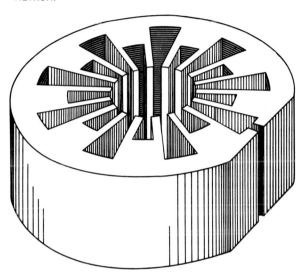

RMA *abbr.* Radio Manufacturers Association.

RMI *abbr.* Radio magnetic indicator.

rms *abbr.* Root mean square.

rock *v.* To move a control back and forth, making other adjustments as necessary, in order to obtain the best alignment.

roentgen *n.* The measure of x- or γ-radiation, defined as the amount such that the associated corpuscular emission per 0.001293 gram of air produces, in air, ions carrying one ELECTROSTATIC UNIT of charge of either sign.

roentgenography *n.* The production of projected images of bodies by means of x-rays.—**roentgenogram** *n.*

roentgenology *n.* The study of x-rays.

Roentgen rays X-RAYS.

roll *n.* The movement of a television picture up or down due to faulty vertical synchronization.

rolloff *n.* An attenuation that varies with frequency, generally increasing at a constant rate beyond the CORNERS of the amplitude-frequency characteristic of a system.

root *n.* **1** Any value of the independent variable in an equation for which the equation is transformed into an identity. **2** Of an arbitrary number *k*, any root of the equation

$$x^n = k,$$

where *n* is a positive integer, called for *n* = 2, 3 the square root and cube root, respectively. When there is a multiplicity of roots it is often implied that the real, positive root is to be chosen.

rooter *n.* A nonlinear AMPLIFIER in which negative feedback is used to make the output voltage proportional to a ROOT of the input voltage.

root mean square See AVERAGE.

rope *n.* A material similar to CHAFF but made in longer strips, dropped from aircraft in the attempt to confuse enemy radars.

rotary *adj.* Characterized by movement about an axis.

rotary switch A switch operated by a rotating shaft.

rotator *n.* **1** In a waveguide, a device changing the plane of polarization, in rectangular waveguides, by twisting the guide itself. **2** A machine for directing an antenna in a desired direction.

rotor *n.* The rotating part of a machine. See STATOR.

round or **round-off** *v.* To adjust (a number) to a lower order of precision by dropping one or more of the least significant digits and compensating for their absence in some way. In one method 5 is added to the digit following the last digit to be retained and a 1 is carried if it occurs. Thus 2.7345 would become 2.735, while 2.7344 would become 2.734.

routine *n.* A computer program designed to carry out some well defined function.

rpm *abbr.* Revolutions per minute.

rps *abbr.* Revolutions per second.

RTL *abbr.* Resistor-transistor logic.

rubber-model method A method of determining the path of an electron in a given two-dimensional electrode configuration by the creation of a gravitational analog of the problem, in which a rubber sheet is stretched over a scale model of the electrodes in which height is made proportional to negative potential. A small rubber ball whose initial velocity is proportional to that of the electron indicates the path taken.

rubidium (Rb) Element 37; Atomic wt. 85.48; Melting pt. 38.5°C; Boiling pt. 700°C; Spec. grav. (20°C) 1.53; Density (20°C) 1.53 g/cm^3; Valence 1; Heat of fusion (39°C) 6.1 cal/g; Heat of vaporization (688°C) 212 cal/g; Heat capacity (39–126°C) 0.0913 cal/(g)(°C); Thermal conductivity (39°C) 0.07 cal/(sec) (cm^2)(cm) (°C); Viscosity (50°C) 6.26 millipoises; Vapor pressure (294°C) 1mm; Electrical resistivity (50°C) 23.15 × 10^{-6} ohm/cm. Uses are similar to those of CESIUM.

rumble *n.* Low-frequency noise that contaminates an audio signal, often resulting from vibrations of mechanical devices such as turntables.

runaway *n.* A condition in which one of the dynamic variables of a system makes an unintended increase to a level beyond the design limits, often with destructive consequences.

ruthenium (Ru) Element 44; Atomic wt. 101.7; Melting pt. 2450°C; Boiling pt. above 2700°C; Spec. grav. (20°C) 12.2; Hardness 6.5; Valence 3, 4, 6, or 8; Crystal structure Close-packed hexagonal, a = 2.70, c = 4.28 at 20°C; Spec. heat (0°C) 0.057 cal/(g)(°C); Coefficient of linear thermal expansion (20°C) 9.6 × 10^{-6}/°C; Modulus of elasticity 60 × 10^6 psi; Electrical resistivity (20°C) 7.2 × 10^{-6} ohm/cm. Used to harden platinum and palladium for use as electrical contacts.

rutherford *n.* A unit of radioactivity, equal to one million nuclear disintegrations per second.

S

S *symbol* **1** (S) ELASTANCE. **2** (s, s_m) Entropy per atom or molecule. **3** (s) Length of arc. **4** (s) Scattering coefficient. **5** (S,s) Sensitivity of phototube: (s) dynamic, (S) static. **6** (s) Specific surface. **7** (s) Speed. **8** (s) Spin quantum number of electric spin of an electron quantized in units of *h*. **9** (s) A differential operator; $s = d/dt$. **10** (s) Complex or generalized frequency variable; $s = \alpha \pm j\omega$.

S *abbr.* Siemens.

sabin *n.* A unit of sound absorption equivalent to one square foot of a completely absorbing substance.

salient pole In an electric motor or generator, a magnetic field pole projecting towards the armature.

samarium (Sm) Element 62; Atomic wt. 150.43; Melting pt. above 1300°C; Spec. grav. 7.7–7.8; Valence 2 or 3; Spec. heat 0.042 cal/(g)(°C); Electrical conductivity (0–20°C) 0.011 $\times$ 10^6 mho/cm.

sampling gate A GATE circuit which transmits or amplifies an input signal only when activated by a selector pulse.

sampling oscilloscope An OSCILLOSCOPE which enables a rapid, periodic waveform to be displayed on a cathode ray tube by stroboscopic sampling of successive waves at different points on the waveform, and representing the sample by a dot on the screen, the accumulation of dots describing the waveform.

sampling theorem In information theory, a theorem stating that a continuous waveform s(*t*) containing frequency components all of which are less than or equal to *f* cycles per second is unambiguously specified by a set of independent sample ordinates spaced *1/2f* second apart.

saturable reactor A magnetic core REACTOR whose reactance is controlled by a varying direct current applied to the core.

saturate *v.* To bring into a state of SATURATION. —**saturable** *adj.*

saturation *n.* **1** A condition in which an increase in an independent variable will make little or no further change in the dependent variable. Thus magnetic saturation is the maximum magnetization of which a body is capable. In anode saturation increases in anode voltage will not result in increases in anode current, **2** In color television, the degree to which a color is mixed with white; the more white, the lower the saturation.

sawtooth *adj.* Of a waveform, increasing approximately linearly as a function of time for a fixed interval, returning to its original state sharply and repeating the process periodically.

saxophone *n.* A linear antenna array having a cosecant-squared radiation pattern.

S band A band of microwave frequencies extending from 1,550 Mhz to 5,200 Mhz.

SBS *abbr.* SILICON BILATERAL SWITCH.

SBT *abbr.* SURFACE BARRIER TRANSISTOR.

SCA *abbr.* SUBSIDIARY COMMUNICATIONS AUTHORIZATION.

scalar *n.* A quantity which is characterized by magnitude alone, unlike a VECTOR, which is characterized by magnitude and direction.

scalar field A region of space in which a parameter, for example temperature, is described at each point by a scalar function.

scalar product An operation on two vectors giving a scalar quantity as a result. The scalar product $\mathbf{A} \cdot \mathbf{B}$ is defined as $|\mathbf{A}|\,|\mathbf{B}|\cos\theta$, where θ is the angle between the two vectors when drawn from a common origin. In a three-dimensional coordinate system

$$\mathbf{A} \cdot \mathbf{B} = A_x B_x + A_y B_y + A_z B_z.$$

scale *v.* To count PULSES by means of a SCALER.

scaler *n.* A device that produces an output pulse when a predetermined number of input pulses have been received.

scaling circuit A SCALER.

scan *v.* To pass a beam of light or electrons rapidly over every point of a surface (image, etc.) as for television reproduction.

scandium (Sc) Element 21; Atomic wt. 44.96; Melting pt. 1200°C; Boiling pt. 2400°C, calculated; Spec. grav. (10°C) 3.02; Valence 3; Spec. heat 0.13 cal/(g)(°C); Electrical conductivity (0–20°C) 0.015 $\times$ 10^6 mho/cm.

scattering *n.* The diffusion of a beam of photons or other particles as a result of collisions with systems or with other particles; also, the diffusion of a beam of radiant energy as a result of anisotropy of the transmitting medium.

Sc.B.E. *abbr.* Bachelor of Science in Engineering.

S.C.C. *abbr.* Single cotton covered.

sce *abbr.* Single cotton enamel.

schematic diagram A stylized drawing of a circuit in which the various elements are represented by conventional symbols.

Schering bridge An ac bridge used to measure an unknown capacitor and its internal series resistance in terms of known resistances and capacitances.

Schmitt trigger A cathode- or emitter-coupled BISTABLE MULTIVIBRATOR. The circuit assumes its 'on' condition for $v_{in} > v_1$ and reverts to its 'off' condition for $v_{in} < v_2$. Since, in general, $v_1 > v_2$, the circuit exhibits HYSTERESIS.

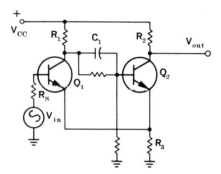

A transistor Schmitt trigger

Schottky effect An increase in the thermionic emission of a conductor as a result of the effectiveness of an electric field on the conductor surface in reducing the WORK FUNCTION of the material.

scintillate v. To emit flashes of light.

scintillation n. The flash of light produced by certain crystalline materials when a charged particle is passed through them.

scope n. OSCILLOSCOPE.

scotophor n. A material that absorbs light when bombarded by electrons, used in DARK-TRACE TUBES.

scp abbr. Spherical candle-power.

SCR abbr. SILICON CONTROLLED RECTIFIER.

scramble v. To encode (a transmitted signal) in such a way as to render it incomprehensible to all but specially equipped receivers, generally by inversion and transposition of frequency bands according to a prearranged scheme. **—scrambler** n.

screen n. 1 The surface of a cathode ray tube on which a visible image appears. 2 A metal partition used to cut off electric or magnetic fields. 3 A SCREEN GRID.

screen grid A GRID placed between the control grid and anode of an electron tube, kept at a fixed positive potential to reduce the electrostatic effect of the anode.

SCS abbr. SILICON CONTROLLED SWITCH.

search radar A radar system designed for giving early warning of the presence of enemy planes or missiles.

sec abbr. 1 SECOND. 2 SECANT.

second n. A unit of time defined as one sixtieth part of a minute, or, more exactly, as $1/31,556,925.9747$ of the tropical year for December 31, 1899 at 12 ephemeris time; now also based on the time of transition between two energy levels in cesium-133.

second anode ULTOR.

secondary n. In a transformer, a winding from which an output is taken.

secondary cell STORAGE CELL.

secondary electron An electron emitted from a solid as a result of bombardment.

secondary service area The area within which the SKY WAVE from a broadcast station can be received without objectionable interference or variation in strength.

secondary standard A physical standard for measurement used in a particular locale, but checked at intervals against a PRIMARY STANDARD.

second detector In a superheterodyne system, a DEMODULATOR as distinguished from a FREQUENCY CONVERTER.

section n. A FILTER SECTION.

Seebeck effect An effect by which a voltage is produced as a result of heating at a junction of two dissimilar metals.

selective adj. Of a receiver, being able to distinguish between the wavelength desired and other nearby wavelengths. **—selectivity** n.

selector n. A multi-position switch used to change the function performed by a piece of apparatus.

selenium (Se) Element 34; Atomic wt. 78.96; Melting pt. (gray form) 220°C; Boiling pt. 688°C; Spec. grav. (20°C, gray) 4.8; Hardness 2; Valence 2, 4, or 6; Spec. heat 0.84 cal/(g)(°C); Electrical conductivity (0–20°C) 0.08×10^6 mho/cm. Upon exposure to light, electrical conductivity of gray form is increased up to a thousandfold. Used in photoelectric cells and in rectifiers.

selenium cell A photoconductive cell whose active material is selenium.

selenium rectifier A rectifier consisting essentially of a junction between a semiconducting layer of selenium and a layer of cadmium, electrons passing more freely from the cadmium to the selenium than vice versa.

self-healing capacitor A capacitor that repairs itself after breakdown from excessive voltage.

selsyn n. SYNCHRO.

semiconductor n. A crystalline material, as germanium or silicon, whose electrical conductivity is between that of a conductor and an insulator.

semiduplex adj. Indicating a communications channel that is operated with one end DUPLEX and the other SIMPLEX.

Sendust n. A soft magnetic alloy of 85% iron, 5% aluminum, and 10% silicon, having initial and maximum permeabilities of 30,000 and 120,000 gauss/oersted, a coercive force of 0.05 oersted and a Curie point of 500°C.

sense n. For a VECTOR, the determination of which of its terminal points is to be considered its beginning and which its end.

sensitivity n. In an electronic device, the ratio of the output to the input signals. —**sensitive** adj.

sensor n. A transducer designed to produce an electrical output proportional to some time-varying quantity, as temperature, illumination, pressure, etc.

separation n. In a stereophonic music-reproduction system, the degree to which a sound that originated in one channel only is suppressed in the other channel on reproduction.

septum n. A thin metal vane in a waveguide perforated with an appropriately shaped hole.

serial digital computer A DIGITAL COMPUTER that operates serially; that is, by handling digits one at a time.

series n. A connection of circuit elements in such a way that the same current flows through each in turn.

series feed The supplying of dc operating power to a tube or transistor in such a way that the dc source is effectively in series with the load impedance.

serrodyne n. A FREQUENCY CONVERTER in which the local oscillator produces a sawtooth waveform which is used to phase-modulate the input signal, giving, theoretically, perfect frequency conversion.

service area The area effectively served by a particular transmitting antenna.

setting time In a FEEDBACK control system, the time required for an error to be reduced to a specified fraction, usually 2 percent or 5 percent, of its original magnitude.

setup n. In television the ratio between reference black and white levels.

S.G. British abbr. SCREEN GRID.

shadow effect The masking effect of hills, etc., in ultrahigh frequency transmission.

shadow factor The ratio of the electric field strength that would result from propagation over a sphere to that which would result from propagation over a plane, all other factors being the same in both cases.

shadow mask A thin metal screen used in color television picture tubes to direct the beams from each of the three electron guns to the correct phosphor dots on the screen.

Shannon's capacity theorem In information theory, a theorem stating that it is possible to encode a source of messages having an information rate H bits/sec so that its information can be transmitted through a noisy channel with an arbitrarily small frequency of errors, provided that $H \leq C$ bits/sec, where C is called the limiting capacity of the channel. This rate C is known for only a few special cases. For a continuous input signal and additive Gaussian noise C is given by **Shannon's formula,** namely

$$C = B \log_2 (1 + M) \text{ bits/sec,}$$

where B is the signal bandwidth and M the signal-to-noise ratio. For a digital channel whose input consists of a source of equiprobable binary digits the capacity is given by

$$C = 1 + p \log_2 p + (1 - p) \log_2 (1 - p)$$

where p is the probability the output differs from the input, which can, in turn, be related to the signal-to-noise ratio.

sheath *n.* **1** The outer covering of a cable or transmission line. **2** The conductive wall of a waveguide. **3** In a gas discharge, a space-charge region.

sheet grating A grating of thin metal sheets extending longitudinally inside a waveguide for about one wavelength, used to suppress unwanted propagation modes.

shf *abbr.* Super-high frequency.

shield *n.* A device designed to protect a circuit, transmission line, etc., from stray voltages or currents induced by electric or magnetic fields, consisting, in the case of an electric field, of a grounded conductor surrounding the protected object. At high frequencies this will provide magnetic shielding as well. At low frequencies (through the audio range) magnetic shielding is accomplished by surrounding the object with a material of high magnetic permeability.

shift *n.* An operation whereby a number is moved one or more places to the left or right. The number 110, for instance, becomes 1100 if shifted one place left or 11 if shifted one place right. The operation is of considerable use in digital computer operations.

shift register In a digital computer, a REGISTER which can shift digital information to the right or left in order to express a higher or lower value in the scale employed.

SHM *abbr.* Simple harmonic motion.

shock-excite To excite a device, network, etc., by means of an impulse or step-function.

shoot-through A condition in which reverse or zero bias is not applied to a silicon controlled rectifier for a time sufficient to restore it to its forward blocking state, the device remaining indefinitely in its conducting state.

shoran *n.* A radio navigation and surveying system using radar pulses to two fixed ground stations.

short *v.* To SHORT-CIRCUIT.

short circuit An electrical connection of negligible impedance between two terminals of a circuit.

short-circuit *v.* **1** To connect a negligibly small impedance between (a pair or terminals). **2** To cause (a current) to flow into a negligibly small impedance.

shorting contact switch A selector switch which contacts a new circuit before it breaks the one connected at the time.

short wave A radio wave having a frequency above 3,000 khz.

shot effect Noise resembling the patter of small shot, developed in an electron tube as a result of the fluctuating emission of electrons from the cathode.

shunt *n.* A parallel path or branch in an electric network, in particular, an admittance connected across an ammeter to reduce its sensitivity.

shunt feed The application of dc operating voltages to a tube or transistor through an inductor that is effectively in parallel with the signal circuit.

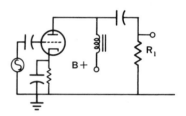

Shunt-fed audio stage

S.I.C., SIC *abbr.* Specific inductive capacity.

sideband *n.* In the spectrum of a modulated carrier, any frequency other than the carrier frequency, that is, any of the frequencies arising as a result of the modulation.

sideband splatter In radio communications, interference on other channels caused by spurious sidebands resulting from overmodulation.

side frequency The frequency of a SIDEBAND.

sidetone *n.* In a telephone or similar system, the sound of his own voice that a speaker hears in his earphone.

sigma *n.* (written Σ, σ) The eighteenth letter of the Greek alphabet, used symbolically to represent any of various coefficients, constants, etc., of which in electronics the principal ones are: **1** (σ) Conductivity. **2** (σ) Dispersion. **3** (σ) STEFAN-BOLTZMANN CONSTANT. **4** (Σ) Summation. **5** (σ) Surface charge density. **6** (σ) Thomson coefficient. **7** (σ) Wave number. **8** (σ) Millisecond.

signal *n.* The intelligence conveyed over a communication system.

signal-to-print ratio In magnetic recording, the ratio of signal to PRINT THROUGH.

signal tracing The process of locating a fault in a circuit by injecting a test signal at the input and checking each stage, usually from the output backwards.

silicon (Si) Element 14; Atomic wt. 28.09; Melting pt. 1420°C; Boiling pt. 2600°C; Spec. grav. (20°C) 2.42; Hardness 7; Valence 4; Spec. heat 0.162 cal/(g)(°C); Electrical conductivity (0–20°C) 0.10×10^6 mho/cm. Silicon is the most abundant electropositive element found in the earth's crust. It has semiconductor properties and is used in rectifiers, transistors, and solar batteries. Silicon dioxide crystals (quartz) are used for piezoelectric crystals.

silicon bilateral switch A semiconductor device which is essentially equivalent to two identical SILICON UNILATERAL SWITCHES connected in inverse parallel, used mainly for triggering TRIACS.

silicon controlled rectifier A three-terminal PNPN DEVICE that maintains a blocking condition between its anode and cathode until a sufficiently large current is made to flow between its gate and cathode. It then breaks into conduction, the magnitude of the current being limited by the external circuit. It remains in conduction until the anode voltage is reduced to zero or less.

silicon controlled switch A four-terminal pnpn semiconductor device that is roughly equivalent to a pnp and an npn transistor connected in a regenerative feedback configuration. Conduction between its anode and cathode may be initiated by a positive pulse at the cathode gate, a negative pulse at the anode gate, or a sufficiently high anode-to-cathode voltage. The device reverts to its blocking condition if the anode voltage becomes zero or negative. In addition, the gate sensitivities are high enough to allow a reverse pulse at either gate to turn the device off.

silicone n. Any of a number of polymers composed of alternate silicon and oxygen atoms with organic radicals attached to the silicon atoms. Their resistance to water, oxidation, and high temperatures make them useful as insulating materials.

silicon unilateral switch A three-terminal pnpn device that breaks into forward conduction when its anode is made sufficiently positive with respect to its cathode.

It is provided with an anode gate through which it may be biased so as to have a higher breakover voltage, used mainly to trigger larger thyristors.

silver (Ag; *Lat.* argentum) Element 47; Atomic wt. 107.880; Melting pt. 960.5°C; Boiling pt. 1950°C; Spec. grav. (20°C) 10.50; Hardness 2.5–2.7; Valence 1; Spec. heat 0.056 cal/(g)(°C); Electrical conductivity (0–20°C) 0.616×10^6 mho/cm. Used in silver solder, in corrosion-resistant batteries, and for electrical contacts. Also used as an emergency substitute for copper in bus bars.

simple harmonic motion See HARMONIC MOTION.

simplex *adj.* Indicating a communications system, in which a single message can be transmitted in a single direction at any one time.

sine n. See TRIGONOMETRIC FUNCTIONS.

sine-cosine potentiometer A special potentiometer having two sliding taps and a specially tapered resistance element arranged so that the voltage between one sliding tap and ground is proportional to the sine of the shaft rotation and that between the other and ground is proportional to the cosine of the shaft rotation, often used as a RESOLVER.

sine wave A wave described by $f(t)$ where

$$f(t) = A \sin (\omega t + \phi).$$

It can be shown that $f(t)$ when formulated in this way includes all linear combinations of $\sin \omega t$ and $\cos \omega t$, dependent only on the values of A and ϕ.

singing n. A parasitic sustained oscillation in or above the passband of a system or component.

single-ended amplifier An amplifier in which each stage uses but a single tube or transistor; more precisely, an amplifier in which each stage is asymmetric with respect to ground.

single sideband transmission An amplitude-modulated radio transmission in which one sideband and the carrier are suppressed.

sink n. Anything into which power of some kind is dissipated.

sinor n. PHASOR.

sinusoid n. Any of the set of curves whose form is the same as that of

$$y = \sin x,$$

including, in fact, all linear combinations of sin x and cos x. —**sinusoidal** adj.

skiatron n.   A DARK-TRACE TUBE.

skin depth   The depth from the surface of a conductor at which the current density drops to 1/e (approximately 37 percent) of its value at the surface.

skin effect   A phenomenon whereby the variation of the current in a conductor causes the current distribution through the cross section of the conductor to become non-uniform, specifically, the tendency (increasing with frequency) of an alternating current to flow in the outer region of a conductor.

sky wave   An IONOSPHERIC WAVE.

slave n.   In a system, a component that does not act independently, but only under the control of another similar component.

slc abbr.   Straight-line capacitance.

sleeping sickness   In an electron-tube BINARY, a condition in which after the circuit remains in one state for an extended time a permanent transition to the other state becomes difficult or impossible as a result of a large interface resistance formed in the tube which has been cut off.

slewing rate or **slew rate**   For an amplifier, programmed power supply, or a related device, the maximum rate of change of the output voltage with respect to time.

S.L.F. British abbr.   Straight-line frequency.

slide switch   A switch actuated by sliding a control lever from one position to another.

slide wire   A device consisting essentially of a resistance wire with a sliding contact that can be fixed to it at any point along its length.

slip rings   Conductive rings by means of which the rotor of an electrical machine maintains contact, through brushes, with the stationary part of the machine.

slope n.   **1** For a straight line, the tangent of the angle included between the line and the positive x-axis. **2** At a point on a curve, the slope of the tangent line drawn to the curve at that point, equal to the derivative of the curve evaluated at that point.

slope detection   A form of fm demodulation in which the incoming signal is applied to a tuned circuit adjusted so that the center frequency of the signal is to one side or the other of resonance. Under these conditions the instantaneous amplitude of the response is roughly proportional to instantaneous signal frequency. The output is thus an amplitude-modulated signal that can be detected in the usual manner. The method gives less than optimum results as the response curve of the tuned circuit is not linear, and any spurious amplitude-modulation of the signal appears at the output.

slot antenna   An antenna in which energy is radiated from a slot cut in a metal surface fed by a coaxial line or waveguide.

slotted section   A section of waveguide or shielded transmission line in which there is a slot to allow the insertion of a test probe.

slug n.   **1** A copper ring encircling the core of a relay, acting as a shorted turn and delaying the operation of the device. **2** In an inductor, a ferromagnetic core that can be moved in order to vary the inductance. **3** In a waveguide, a piece of metal or dielectric material than can be moved in order to effect tuning or impedance matching. **4** A measure of mass, defined as that mass which is accelerated at 1 ft./sec^2 by a force of 1 pound.

slump n.   In a video amplifier tube, a little-understood defect that causes the positive response at the plate resulting from a negative step applied at the grid to droop.

slw abbr.   Straight-line wave length.

small signal   A signal sufficiently small that the EQUIVALENT CIRCUIT of an inherently nonlinear device (as a transistor or vacuum tube) to which it is applied is sufficiently accurate if only linear elements are included. Circuit analysis under these conditions is far simpler than for LARGE SIGNALS.

smear n.   In a television picture, a defect in which objects appear to spread horizontally beyond their usual boundaries as though rubbed, resulting often from defects in the peaking network of the video amplifier.

S meter   A device used in some radio receivers to measure signal strength.

Smith chart   A circular chart on which the real and

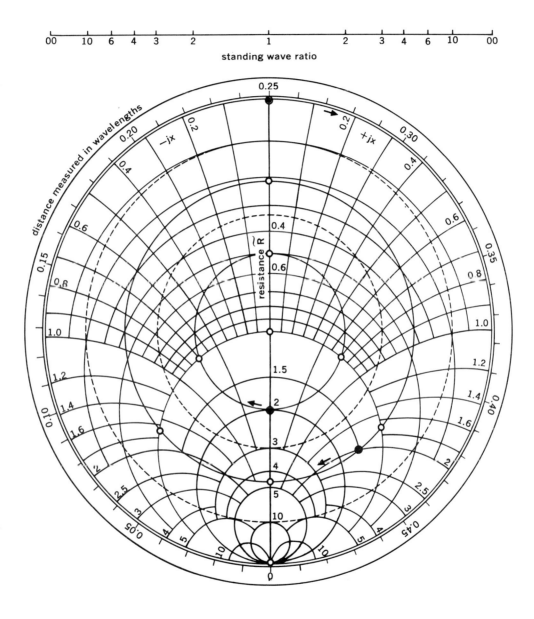

standing wave ratio

distance measured in wavelengths

resistance R

−jx   +jx

[145]

imaginary parts of any normalized IMMITANCE are plotted on coordinates consisting of two families of orthogonally intersecting circles. The chart makes possible rapid graphical solutions to problems dealing with transmission lines and antennas. See page 145.

Smith no. 10   An alloy of 54% Iron, 38% Chromium, 8% Aluminum. High resistance alloy; oxidation-resistant at about 1315°C.

smooth *v.*   To attenuate the high-frequency components of (a waveform), as by filtering or integration.

snap diode   A type of silicon junction diode designed to enter its nonconducting state extremely rapidly when switched from forward to reverse bias.

Snell's laws   Either of a pair of laws describing the behavior of electromagnetic radiation (or with slight alterations other radiant energy as well) as it passes from one medium to another. The first law states that the incident ray, the normal to the surface of the interface at the point of incidence, and the refracted ray all lie in a single plane. The second law states that if i_1 is the angle of incidence and i_2 is the angle of refraction, then

$$\sin i_2 / \sin i_1 = n_1 / n_2,$$

where $n_{1, 2}$ are constants (properly **indices of refraction**) that depend on the media. For a case where $n_1 > n_2$ and $\sin i_1 > n_2/n_1$ the above relation cannot be satisfied and there is total internal reflection. For sound the law takes the form

$$\sin i_2 / \sin i_1 = c_2 / c_1,$$

where $c_{1, 2}$ are the speeds of sound in the two media.

snivet *n.*   In television, a black line, often irregular in form, appearing near the right edge of the screen, caused by a discontinuity in the plate-current characteristic of the horizontal amplifier tube under conditions of zero bias.

snow *n.*   The flecked pattern appearing on a television screen as a result of a poor signal-to-noise ratio.

SNR *abbr.*   Signal-to-noise ratio.

sodar *n.*   A system for determining the temperature and height of various levels of the atmosphere by projecting sound straight up and analyzing the echoes that return.

sodium (Na; *Lat.* natrium.**)**   Element 11; Atomic wt. 22.997; Melting pt. 97.5°C; Boiling pt. 880.0°C; Spec. grav. (20°C) 0.971; Density (0°C) 0.972 g/cm³, (100°C) 0.928 g/cm³, (800°C) 0.757 g/cm³; Hardness 0.4; Heat of fusion (97.5°C) 27.2 cal/g; Heat of vaporization (883°C) 1005 cal/g; Heat capacity (20°C) 0.30 cal/(g)(°C), (200°C) 0.32 cal/(g)(°C); Thermal conductivity (21.2°C) 0.317 cal/(sec)(cm²)(cm)(°C); Viscosity (250°C) 3.81 millipoises, (400°C) 2.69 millipoises; Vapor pressure (440°C) 1 mm, (815°C) 400 mm; Surface tension (100°C) 206.4 dynes/cm, (250°C) 199.5 dynes/cm; Electrical resistivity (100°C) 965 × 10⁻⁶ ohm/cm. Liquid sodium metal is used as a heat-transfer agent in nuclear reactors.

soft *adj.*   **1** Designating an electron tube that contains gas, either by design or because its vacuum has deteriorated. **2** Designating x-rays of relatively little penetrating power.

soft solder   An alloy of tin and lead, used for soldering. The eutectic with 63% tin freezes sharply at 183°C.

software *n.*   Any of the specially prepared programs designed to optimize the operation of a digital computer, including programmer aids such as assemblers, processors, etc., and more complex programs such as operating systems that to a large extent manage the entire installation.

solar cell   A transducer, generally of low efficiency, designed to convert light into electric power.

solder *n.*   An alloy with a low melting point used in joining metals whose melting point is much higher.

solenoid *n.*   A coil of wire surrounding a movable iron bar that is located in such a way that when the coil is energized the core is drawn into it.

solid angle   The figure formed by all the lines that pass through a given point and a simple closed curve.

solion *n.*   An electrochemical amplifier in which four electrodes are immersed in an electrolyte, the current passing between one pair controlling the conductivity with respect to the other pair. Typically the device acts as an integrator.

sonar *n.* A system of underwater detection and ranging, similar in operation to radar but using ultrasonic sound waves rather than radio waves.

sone *n.* The measure of sound intensity, equivalent to a simple tone having a frequency of 1,000 cycles per second at 40 decibels above the listener's audibility threshold.

sonic *adj.* Of or having to do with sound.

sonne *n.* A radionavigation system having three antennas at distances of 2.88 wavelengths and using phase differences for determination of bearings.

sonobuoy *n.* A floating sonar device used by aircraft for the detection of submarines. Several sonobuoys broadcasting on different wavelengths are commonly used.

sonoluminescence *n.* LUMINESCENCE caused by sound.

sound *n.* A mechanical disturbance in an elastic medium. Sound is audible at frequencies from about 15 to 20,000 hz.

sound-on-sound The process of playing back an existing recording, mixing it with a simultaneous live sound or with another recording, and recording the mixture. See SOUND-WITH-SOUND.

sound pressure Pressure caused by sound defined as the difference between the actual pressure at a particular point and the static pressure of the medium.

sound-with-sound The synchronized playback of what are essentially two (or more) different recordings.

source *n.* **1** Anything that supplies energy. **2** In a field effect transistor, the terminal analogous with the cathode of a vacuum tube.

south pole The end of a magnet into which magnetic lines of force are said to flow. See LEFT HAND RULE.

SP *abbr.* Single pole.

space charge The electric charge resulting from a large population of particles, having charges of the same polarity and distributed through a volume of space.

space charge region DEPLETION REGION.

spacistor *n.* A transistor in which carriers are injected into the space-charge-free layer of a *p-n* junction,

having a low transient time and capable of operating at frequencies up to 10,000 Mhz.

spade bolt A bolt with a thread at one end and flat section with a hole through it at the other.

spade tip A terminal connector consisting of a flat piece of metal with a hole in it attached to a wire.

spaghetti *n.* A form of tubular insulation that can be slipped over wires before they are connected to terminals.

spark *n.* The luminous effect of a disruptive electric discharge or the discharge itself.

spark gap The region between and over two electrodes where a SPARK may occur. With spherical electrodes 5cm in diameter 2 cm apart a spark will correspond to a potential difference of 56,300 volts in air at atmospheric pressure. Generally the sparking potential is a linear function of the gas pressure times the distance between the electrodes.

spdt *abbr.* Single-pole double-throw.

speaker *n.* A LOUDSPEAKER.

spec *informal* Specification.

specific gravity The ratio of a mass of a body to that of an equal volume of some standard substance, water in the case of solids, and liquids and air or hydrogen in the case of gases.

specific heat The amount of heat required to raise one gram of a specified substance 1°C.

spectro- *combining form* Radiant energy; as exhibited in the spectrum.

spectroscope *n.* A device to measure the properties of a spectrum by dispersing radiant energy according to wavelength by means of a prism or diffraction grating. See also MASS SPECTROSCOPE.

spectrum *n.* An array of physical entities arranged in order of the value of some parameter, in particular, a set of electromagnetic waves arranged according to frequency. The electromagnetic spectrum may itself be divided into subspectra, as the radio spectrum, the visible spectrum, etc. —**spectral** *adj.*

speech clipper In radio communications; a device that limits the peaks of speech waveforms, allowing an increase of the average modulation of the carrier.

speedlight *n.* STROBOSCOPE.

sp. gr. *abbr.* Specific gravity.

sp. ht. *abbr.* Specific heat.

spider *n.* A flexible, perforated disk serving to keep the moving coil of a dynamic microphone or loudspeaker centered in its magnetic gap without restricting its longitudinal motion.

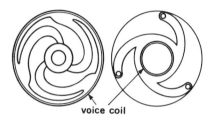

voice coil

spill *n.* In a STORAGE TUBE, loss of information caused by redistribution of secondary electrons.

spkr *abbr.* Loudspeaker.

s-plane A COMPLEX PLANE on which are plotted the real and imaginary parts of $s = -\alpha \pm j\omega$. See COMPLEX EXPONENTIAL.

s.p.s.t. *abbr.* Single pole, single throw.

sputtering *n.* The ejection of atoms from the metal cathode of a tube due to bombardment by positive ions. The effect can be used to form a fine coating on a surface deliberately exposed to the ejected atoms.

square loop material A ferromagnetic material that has an approximately square or rectangular HYSTERESIS loop, used in the construction of various bistable elements as in computer core storages, etc.

square wave A periodic waveform consisting of a train of equispaced rectangular PULSES having a DUTY FACTOR of 50 percent.

squeal *n.* In a radio receiver, a high pitched tone heard together with the wanted signal.

squegg *v.* To oscillate and cut off for alternate periods of time, especially as a result of grid blocking, as in an oscillator circuit. **—squegger** *n.*

squelch **1** *v.* To disable the audio circuits of (a radio receiver) when there is no signal of interest or when it is tuned between channels, often by automatic means. **2** *n.* A device that accomplishes this.

squitter *n.* In a radar transponder, excitation caused by random noise.

SSB *abbr.* Single sideband.

ssc *abbr.* Single silk-covered.

ST *abbr.* Single throw.

stability *n.* In a feedback amplifier or control system, freedom from undesired oscillation, in particular, for a linear system, a property whereby its response to any bounded input is also bounded. **—stable** *adj.*

stabistor *n.* A switching diode having closely controlled voltage-current characteristics.

stable equilibrium See EQUILIBRIUM.

stack *n.* An array of similar or identical components.

stage *n.* One of the functional parts into which a circuit can be conveniently divided, containing In general one or more active devices.

stagger tuning In a tuned cascade amplifier, the use of different resonant frequencies at each stage to provide broader overall bandwidth.

staircase wave A waveform consisting of a series of step functions all having the same polarity.

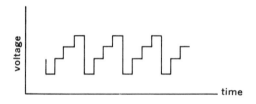

stalo *n.* An extremely stable local oscillator used to provide a reference signal in moving target radar, enabling stationary objects to be eliminated from the display.

stamper *n.* The negative image of a phonograph record, used to press copies in vinyl.

standing wave A wave disturbance resulting from the superposition of two waves of equal frequency and amplitude traveling through a medium in opposite directions. For two such sinusoidal waves the sum $s(x,t)$ is given by

$$s(x, t) = 2A \cos (2\pi x/\lambda) \cos \omega t,$$

where A is the amplitude of the original waves, x is position, ω is the radian frequency, and λ the wavelength. The amplitude of a standing wave is a function of position and time, but as a result of fixed phase at every point no net propagation occurs.

standing-wave ratio   In a transmission line, waveguide, or analogous system, a figure of merit used to express the efficiency of the system in transmitting power, specifically taking into account the effects of MISMATCH between source, line, and load. The standing wave ratio S is given by

$$S = V_{max}/V_{min} = I_{max}/I_{min}.$$

In an ideally matched system $S = 1$, indicating the presence of a pure traveling wave and no reflected power. As the proportion of the power reflected increases S approaches infinity, which condition would indicate a pure standing wave.

standoff *n.*   An insulator designed to keep a conductor at a distance from the surface that supports the insulator.

Stark effect   The splitting of a line in the spectrum of a radiating atom into several components when the atom is subjected to an electric field.

star network   A network consisting of three or more branches each of which has one terminal connected to a common node.

starter *n.*   In a glow discharge tube, an electrode used to ionize the gas and thus initiate conduction.

starved amplifier   A pentode stage in which the screen grid voltage is unusually low and the plate load resistance is unusually high, giving very high stage gain at low current drain, but often at the expense of frequency response.

-stat *combining form*   A device that stops or makes constant, as THERMOSTAT, RHEOSTAT.

statampere   The measure of current in the electrostatic cgs system, equal to 3.3356×10^{-10} ampere (absolute).

statcoulomb   The measure of electric charge in the electrostatic cgs system, equal to 3.3356×10^{-10} coulomb (absolute).

statfarad   The measure of capitance in the electrostatic cgs system, equivalent to 1.11263×10^{-12} farad (absolute).

stathenry   The measure of inductance in the electrostatic cgs system, equivalent to 8.98766×10^{11} henrys (absolute).

static *n.*   ATMOSPHERICS.

static charge   A stationary electric charge.

staticizer *n.*   In a computer, a storage device that can accept information sequentially and reproduce it in parallel.

static machine   A generator producing static electricity by induction.

static switch   A SWITCH using no moving parts.

statistical communication theory   The study, on a statistical basis, of the relation between the amount of information transmitted through a communications channel and the amount actually received, dealing particularly with the effects of noise and of various methods of coding and detection.

statmho   The measure of conductance in the electrostatic cgs system, equal to 1.1126×10^{-12} mho (absolute).

statohm   The measure of resistance in the electrostatic cgs system, equal to 8.98766×10^{11} ohm (absolute).

stator *n.*   In a rotary electric machine, the part that does not move.

statvolt   The measure of electromotive force in the electrostatic cgs system, equal to 299.796 volts (absolute).

stave *n.*   Any of the longitudinal elements that make up a sonar transducer.

STC *abbr.*   Sensitivity time control.

std *abbr.*   Standard.

steady-state *adj.*   Indicating a circuit response that is either periodic or invariant with respect to time.

steatite *n.*   A dense ceramic material composed chiefly of a magnesium silicate, useful for its excellent insulating properties.

steer *v.*   To adjust the polar response pattern of (an antenna) by electrical means.

steerable antenna   A directional antenna.

Stefan-Boltzmann law   A law stating that the total radiation E from a BLACK BODY is given by

$$E = \sigma T^4,$$

where T is the absolute temperature of the body and σ is the **Stefan-Boltzmann constant,** equal to 5.672×10^{-5} erg/ sec cm^2 deg^4.

Steinmetz's formula An approximate formula for the HYSTERESIS loss (L) per cycle when a given sample is magnetized and demagnetized, namely

$$L = aB_m^{1.6},$$

where B_m is the magnitude of the maximum induction and a, the **Steinmetz coefficient,** is constant for any given material.

stenode circuit An intermediate frequency circuit that achieves high selectivity by the inclusion of a piezoelectric crystal in the tuning network of at least one stage of amplification.

step-down adj. Designating a transformer whose output voltage is lower than its input voltage.

step function A function defined by the following equations:

$$f(t) = A, t_0 < t; f(t) = B, t > t_0,$$

where A, B are constants. The **unit step function** $U(t)$ appears often in the analysis of electric networks. For $U(t)$ $A = 0$, $t_0 = 0$, $B = 1$. Like any discontinuous function the unit step is undefined at the point of discontinuity. See DELTA FUNCTION. As it is possible to approximate any function by a sum of step functions, the response of a circuit to a step function is a convenient generalization of its behavior.

step-up adj. Designating a transformer whose output voltage is higher than its input voltage.

steradian n. A measure of SOLID ANGLES, defined as equal to that solid angle which when its vertex is at the center of a sphere subtends an area equal to the square of its radius.

stereocephaloid microphone A microphone designed to simulate the human head, often used in stereophonic recording.

stereophonic adj. Designating a sound reproduction system in which sound is delivered to the listener through at least two channels, creating the illusions of depth and of locality of sources. —**stereophony** n.

stiction n. Static friction.

stiff adj. **1** Designating a voltage source whose value is largely independent of the current drawn, that is, a source of relatively low impedance. **2** Having a high degree of STIFFNESS.

stiffness n. The reciprocal of compliance. See ACOUSTIC STIFFNESS.

stilb n. A measure of luminance equivalent to one candle per cm².

stochastic adj. Random.

stoichiometry n. The calculation and interpretation of mass and energy relationships in chemical reactions, making use of measured quantities of the materials involved and the heat involved in the reaction.

stop band The range of frequencies in which a FILTER, network, transducer, etc., has a high attenuation.

storage n. In a digital computer, any of the devices used to store information for later use, including such hardware as bistable solid-state circuits, magnetic cores, magnetic tapes, etc.

storage battery A group of storage cells.

storage cell A cell that can be recharged by reversing the current.

storage counter A counter in which a series of current pulses are allowed to charge a capacitor, each pulse raising the voltage to a higher level. A comparator determines when the capacitor voltage reaches a predetermined level. Various techniques are often used to linearize the charging curve of the capacitor.

storage location In the STORAGE of a digital computer, a position, generally with a specific ADDRESS, capable of accommodating one WORD.

storage time In a transistor, the interval between the time when an input has fallen 10 percent from its maximum amplitude and the time when the output pulse has fallen proportionately.

storage tube An electron tube into which information can be stored and later read out, usually a cathode ray tube with a storage screen which will retain charges impressed on it and which can control an electron beam in some way, allowing changes to be read out.

S.T.P., stp abbr. Standard temperature and pressure.

strain n. A measure of deformation in a material, caused by stress. Strain can be linear, shear, or volumetric.

strain anisotropy A property of a material whereby a strain developed in it as a result of a stress depends not only on the magnitude of the stress but also on its direction.

strap *n.* In a multi-cavity MAGNETRON, a piece of metal used to connect alternate segments to separate the modes.

strays *n. pl.* Electromagnetic waves affecting a receiver, produced by atmospheric electric discharges and electrical storms.

streaking *n.* The lateral distortion at the edge of a television picture tube when there is a large change from black to white or the reverse on opposite vertical edges.

stress *n.* A force or system of forces that tends to induce deformation in a body.

strip chart recorder An instrument recording variations in the measurement of a quantity by time, using a moving pen on a long strip of paper.

strobe pulse The short flash given out by a STROBOTRON

stroboscope *n.* An instrument for making rapid motion visible by illuminating it for times sufficiently short to make it appear stationary, usually by means of a gas discharge tube.

strobotron *n.* A gas discharge tube used in a STROBOSCOPE.

strontium (Sr) Element 38; Atomic wt. 87.63; Melting pt. 800 °C; Boiling pt. 1150°C; Spec. grav. (20°C) 2.54; Density 2.6 g/cm³; Valence 2; Electron configuration 1 8 18 18 8 2; Ionic radius 1.13Å; Atomic vol. 34.5 cm³/g-atom, Latent heat of vaporization (1150°C) 383 kilojoules/g-atom; Spec. heat 0.176 cal/(g)(°C); Electrical conductivity (0–20°C) 0.043 × 10⁶ mho/cm.

stub *n.* A short-circuited section of transmission line connected across the conductors of a main transmission line and adjusted in length so as to provide an optimum impedance match.

stylus *n.* The part of a phonograph pickup that actually makes contact with the record groove.

subcarrier *n.* A modulated CARRIER which is used in turn to modulate another CARRIER.

subchassis *n.* The supporting base of a MODULE or SUBASSEMBLY.

subcycle generator In a telephone system, a circuit that reduces the frequency of the ringing current to a submultiple, used to provide selective ringing for subscribers sharing a common line.

sub-routine In a digital computer, a set of instructions that perform one operation, as the extraction of a square root, and which can be called upon repeatedly throughout a program.

Subsidiary Communications Authorization An arrangement whereby the unused bandwidth in an FM stereo channel is allocated, through additional frequency division multiplexing, to other services, in particular to transmission of commercial background music.

substrate *n.* The support for a microelectronic device, often used as a part of the circuit or device.

suckout *n.* An undesired dip in response anywhere within the passband of a device.

Suhl effect An effect whereby holes injected into a filament of n-type semiconductor material are deflected by a strong transverse magnetic field to the surface, where they may speedily recombine or be withdrawn by a probe.

sulfur (S) Element 16; Atomic wt. 32.006; Melting pt. (rhombic) 112.8°C, (monoclinic) 119.0°C; Boiling pt. 444.6°C; Spec. grav. (20°C) (rhombic) 2.07, (monoclinic) 1.957; Valence 2, 4, or 6; Spec. heat 0.175 cal/(g)(°C); Electrical conductivity (0–20°C) 10⁻¹⁷ mho/cm. Used in certain types of electrical insulators.

sum frequency In a FREQUENCY CONVERTER or other nonlinear transducer, an output frequency equal to the sum of two input frequencies.

summer *n.* In an ANALOG COMPUTER, a component de-

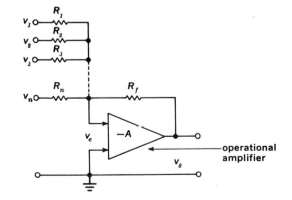

signed to add several continuously varying voltages. It consists of an OPERATIONAL AMPLIFIER in conjunction with a resistive network.

super- *prefix* Above; beyond; over.

superconductivity *n.* The property, exhibited by certain non-ferromagnetic metals ond alloys, of becoming completely non-resistive and diamagnetic when cooled to some transition temperature close to $0°$K. —**superconductive** *adj.*

superconductor *n.* Any metal or compound possessing SUPERCONDUCTIVITY.

superhet *n.* A superheterodyne radio receiver.

superheterodyne *n.* A radio receiver in which incoming signals are converted to a fixed frequency (called the intermediate frequency) where all or most of the amplification takes place, tuning of the receiver being accomplished by varying the frequency of the local oscillator whose output heterodynes with the incoming signal.

Supermalloy *n.* A soft magnetic alloy of 16% iron, 5% molybdenum, and 79% nickel, having initial and maximum permeabilities of 100,000 and 1,000,000 gauss /oersted, a coercive force of 0.002 oersted and a Curie point of $400°C$.

superposed circuit An additional channel that uses the circuitry provided for other channels without interfering with them in any way.

superposition theorem A theorem stating that the response of a linear system to an excitation consisting of several components is equal to the sum of the responses to each of those components taken individually.

superregenerative receiver A radio receiver using a radio-frequency amplifier whose gain is increased to the maximum by the application of sufficient positive feedback to cause oscillation, the oscillations being periodically quenched due to the presence of a radio-frequency component of lower frequency in the plate supply.

supersync signal In television, a combined vertical and horizontal SYNC SIGNAL.

superturnstile antenna A TURNSTILE ANTENNA with wing-shaped dipoles for omnidirectional radiation.

suppressor *n.* 1 An element or device used to suppress an unwanted effect. 2 A SUPPRESSOR GRID.

suppressor grid In a pentode, an electrode placed between the plate and screen grid and maintained at a potential very much less than that of either in order to prevent the transfer of secondary electrons from one to the other.

surface-barrier transistor A transistor in which the collector and emitter are etched or plated on opposite sides of a semiconductor wafer.

surface noise In a music reproduction system, noise arising from friction between the playback stylus and the record groove.

surface reflection The reflection of a part of the incident radiation from the surface of a refractive material.

surge *n.* A large-amplitude pulse of current or voltage.

surge impedance CHARACTERISTIC IMPEDANCE.

SUS *abbr.* SILICON UNILATERAL SWITCH.

susceptance *n.* The imaginary part of ADMITTANCE; the reciprocal of REACTANCE.

susceptibility *n.* 1 In a dielectric material, the measure χ_e of its ease of polarization, given by

$$\chi_e = P/\varepsilon_0 E,$$

where **P** is polarization, **E** electric field strength, and ε the permittivity of free space. 2 In a magnetic material, the measure χ_m of its ease of magnetization, given by

$$\chi_m = M/H,$$

where **M** is magnetization and **H** magnetic field strength.

SW *abbr.* 1 SWITCH. 2 SHORT WAVE.

swamping resistor A resistor connected in series with the emitter of a transistor so as to make the nonlinear and temperature-dependent resistance of the emitter a negligible fraction of the total.

sweep 1 *n.* The horizontal movement of the electron beam in a cathode-ray tube. 2 *n.* A sawtooth waveform, in particular one used to move the electron beam horizontally across the face of a cathode-ray tube. 3 *n.* Anything that changes steadily with time. 4 *v.* To change or cause to change steadily with time.

sweep generator A signal generator whose output frequency is swept back and forth between two limits.

S.W.G. *abbr.* Standard Wire Gauge.

swing *n.* The total variation in frequency or amplitude of a quantity.

swinging choke A filter inductor designed with an air gap in its magnetic circuit so that its inductance decreases as the current through it increases. Used in a power-supply filter, a swinging choke can maintain approximately CRITICAL INDUCTANCE with wide variation in load current.

switch 1 *n.* A circuit element (especially one that is mechanical in nature) having two conditions, one in which its impedance is virtually zero, and another in which its impedance is virtually infinite. Switches are often arranged so that many are actuated by a single control. **2** *v.* To control by means of a switch. **3** *v.* To change from one state to another.

switchboard *n.* A panel or arrangement of panels bearing switches for connecting and disconnecting electric circuits.

switching algebra BOOLEAN ALGEBRA, when applied to switching circuits, digital systems, etc.

switching characteristics Those characteristics of a device, especially of a transistor, diode, or vacuum tube, which bear most heavily on its efficiency when used in switching mode.

switching coefficient For a bistable magnetic element, the derivative of magnetizing force with respect to the reciprocal of the switching time that results.

switching mode A method of utilizing a vacuum tube or transistor so that except for negligibly small transition times it is either in cutoff or saturation. A transistor operated in this way can switch quite large currents with little power dissipation.

switching theory The theory dealing with circuit elements that have two or more discrete states; the theory of digital systems.

SWL *abbr.* Short-wave listener.

SWR *abbr.* Standing-wave radio.

symmetry *n.* A property of an object or configuration of objects whereby it is transformed into itself under a particular operation, that is, if an object K is transformed by an operation T in such a way that $T(K)$ is congruent with K, then K is said to possess symmetry with respect to T. In most of the familiar, elementary cases T is a rotation, a reflection, or a combination of these. The mathematical theory of groups provides powerful insights into the nature of symmetry. Functions and waveforms often possess symmetry; for instance if $y = f(x) = f(-x)$, f is said to be symmetrical with respect to the y-axis.

sync *n., v.* SYNCHRONIZATION; SYNCHRONIZE.

synchro *n.* Any of various transducers whose electrical outputs vary in magnitude and/or phase according to the angular position of a rotor, or whose rotor positions vary according to the magnitude and/or phase of their electrical inputs, used extensively in servomechanisms.

synchronism *n.* The condition of being synchronous.

synchronization *n.* The keeping of one operation in step with another.

synchronize *v.* To cause to operate in unison.

synchronous *adj.* Having the same frequency; occurring at the same time.

synchroscope *n.* **1** An instrument for determining the frequency and phase error between two periodic quantities. **2** An oscilloscope that displays short-duration pulses by means of a fast, synchronized sweep.

synchrotron *n.* A particle accelerator, a modification of the BEVATRON using an alternating electric field synchronized with the orbital motion of the particle. Synchrotrons can accelerate both electrons and heavier particles.

sync signal A signal that serves to keep two systems in synchronization, as in television, where between segments of video information pulses that keep the oscillators generating the horizontal and vertical sweeps in synchronism with those at the station are sent.

synthesizer *n.* In electronic music, a device that produces an audio signal equivalent to a musical composition directly from the composer's specifications and without the intervention of a performer.

system *n.* A group of components so interconnected as to perform a specific function.

T

t *abbr.* Temperature.

T *symbol* **1** (*T*) Temperature K. **2** (*t*) Time. **3** (*T*, $t_{1/2}$) Radioactive half-life. **4** (*T*) Oscillation period.

tachometer *n.* An instrument for measuring speed or velocity, and, in particular, angular velocity.

tandem *n.* CASCADE.

tandem transistor See COMPOUND CONNECTION.

tank circuit A resonant *LC* circuit.

tantalum (Ta) Element 73; Atomic wt. 180.88; Melting pt. 2996°C; Boiling pt. above 4100°C; Spec. grav. 16.6; Valence 3 or 5; Spec. heat 0.036 cal/(g)(°C); Electrical conductivity (0–20°C) 0.081 $\times$ 10⁶ mho/cm. Used as a resistant electrode, in electric current rectifiers, in radio and radar tubes, and in electrolytic capacitors. Was previously used as a filament for the incandescent electric lamp, but has been replaced by tungsten.

tap *n.* An electrical connection made to a resistor or coil at a point along its length.

tape deck A tape recorder having no internal loudspeakers or provision for power amplification.

tape loop An endless band of recording tape used for repeating messages.

taper *n.* A function relating the angular position of a shaft to the resistance of a rheostat or voltage-division ratio of a potentiometer. For instance, an audio gain control might be designed so that

$$V_{out}/V_{in} = 10^{(-\phi/20)},$$

where $\phi = 0$ is the extreme right position of the shaft, so that a change $\Delta\phi$ always represents a constant number of decibels.

tape recorder A device that converts sound into magnetic patterns stored on tape, reversing the process for playback.

target *n.* **1** Anything exposed to radiation. **2** The anode of an x-ray tube. **3** Anything reflecting a sonar or radar beam. **4** In an IMAGE ORTHICON, the signal electrode.

tau (*written* T, τ) The nineteenth letter of the Greek alphabet, used symbolically to represent any of various constants, coefficients, etc., of which in electronics the principal ones are: **1** (τ) Time **2** (τ) Transmittance **3** (τ) Decay modulus.

taut-band galvanometer A GALVANOMETER whose moving coil is suspended between two taut ribbons.

TE *abbr.* Transverse electric. See MODE.

tear *n.* A defect in a television picture caused by faulty horizontal synchronization.

teasing *n.* The repeated slow opening and closing of a set of CONTACTS in an energized circuit in order to test their useful lifetime.

technician license A license issuable by the FCC enabling a ham to operate in the frequency range 50.25 to 54 Mhz, 145 to 147 Mhz and above 220 Mhz after Nov. 22, 1969. Its requirements include the ability to send and receive code at 5 words per minute and a knowledge of general radio theory and practice.

technitium (Tc) Element 43; Atomic wt. 99; Valence 6 and 7 observed. Radioactive with a half-life of 2 $\times$ 10⁵ years.

tecnetron *n.* A semiconductor device with cathode and anode connections at each end of a rod of germanium which is surrounded by an indium ring controlling carrier flow through the germanium by the field effect.

tee *n.* A network consisting of three branches meeting at a common node; an alternate form of WYE.

tee junction A waveguide junction in the form of a T.

Teflon *n.* The trade name of a polymerized tetrafluoroethylene which has high chemical resistance and can withstand temperatures over 200°C, used as an insulator and dielectric.

telautograph *n.* A device for transmitting an instantaneous copy of writing or drawing.

telegraph *n.* A system of transmitting messages by electrical means using a code.—**telegraphy** *n.*

telemeter *v.* To measure (a quantity) from a distance.

telephone *n.* A system for transmitting speech or other sounds over a communication channel between individuals.—**telephony** *n.*

telephoto *n.* FACSIMILE.

teleprinter *n.* A device for sending messages in type-written form over a telegraph circuit.

teletypewriter *n.* TELEPRINTER.

television *n.* The transmission of continuous visual images as a series of electrical impulses or a modulated carrier wave, the signals being restored to visual form on the cathode-ray screen of a receiver, usually with accompanying sound.—**televise** *v.*

telex *n.* A subscription teleprinter service similar in many respects to the telephone service.

tellurium (Te) Element 52; Atomic wt. 127.61; Melting pt. 452°C; Boiling pt. 1390°C; Spec. grav. (20°C) 6.24; Hardness 2.3; Valence 2, 4, or 6; Spec. heat 0.47 cal/(g)(°C); Electrical conductivity (0–20°C) 10 mho/cm. Used as a brightener in electroplating baths; certain alloys have high electrical resistance.

TEM *abbr.* Transverse electromagnetic. See MODE.

tera- *combining form* A trillion (10^{12}) times a specified unit.

terbium (Tb) Element 65; Atomic wt. 159.2; Melting pt. 1356°C; Boiling pt. 2800°C; Density 8.27 g/ml; Valence 3; Spec. heat 0.044 cal/(g)(°C); Electrical conductivity (0–20°C) 0.009×10^6 mho/cm.

terminal *n.* A point at which a network or component can be connected to another network or component.

termination *n.* The load connected to the output of a device or transmission line.

tesla *n.* A unit of magnetic induction equivalent to one weber per square meter.

Tesla coil An INDUCTION COIL whose primary has a high-frequency spark gap instead of an interrupter, inducing in the secondary voltage sufficient to cause an intense high-frequency discharge.

test pattern A geometric pattern used to test the image fidelity of a television system.

test record A disk containing various accurately known signals with which to test phonographs.

test set An assembly of instruments and circuit elements for testing a particular device.

tetrode *n.* A four terminal active device, in particular, a vacuum tube containing a cathode, control grid, screen grid, and plate.

tetrode transistor A bipolar transistor having an addi-tional contact to the base region. Through this contact a reverse bias is applied to a portion of the base region, reducing its effective area and junction capacitance, thus extending the frequency response of the device.

thallium (Tl) Element 81; Atomic wt. 204.39; Melting pt. 303.5°C; Boiling pt. 1650°C; Spec. grav. (20°C) 11.85; Valence 1 or 3; Valence-electron configuration $6s^2 6p^1$; Spec. heat 0.031 cal/(g)(°C); Electrical conductivity (0–20°C) 0.055×10^6 mho/cm.

thalofide cell A photoelectric cell in which the active material is thallium oxysulphide, sensitive to red and infrared light.

thermal agitation noise Noise produced by the random movement of current carriers in a conductor. The mean square noise voltage is given by

$$v^2 = 5.49 \times 10^{-23}\ TR\Delta f$$

where T is absolute temperature, R the resistance of the conductor, and Δf the bandwidth.

thermal converter A device for the production of electricity from heat directly by the SEEBECK EFFECT.

thermal noise The random noise caused by the thermal excitation of molecules, atoms, or particles; in particular THERMAL AGITATION NOISE.

thermal relay A relay whose switching element is actuated by an electrically heated device rather than by electromagnetic action, operating, in general, with a predictable delay between the start of the energizing current and the switching response.

thermal runaway In a transistor, a regenerative condi-tion in which heating at the collector junction causes collector current to increase, which in turn causes more heating, etc., the temperature rapidly approaching levels that are destructive to the transistor.

thermel *n.* A thermocouple or series of thermocouples used to measure temperature.

thermion *n.* An electrically charged particle emitted by a heated body.

thermionic emission The emission of ions or electrons from a metal as the result of heating. The thermionic current density, J, in amperes/cm^2 is equal to $AT^2 e^{-\phi/kT}$ where A is a constant, T is degrees Kelvin, k is BOLTZMANN'S CONSTANT, and ϕ is the WORK FUNCTION of the emitter.

thermo- *combining form*   Heat.

thermocouple *n.*   A thermoelectric device using the principle that when two dissimilar metals are welded together and the junction heated a voltage will be developed across it.

thermojunction *n.*   The point of contact between the two conductors of a THERMOCOUPLE.

thermophone *n.*   An electroacoustic transducer that produces sound waves as a result of the expansion and contraction of air around a conductor whose heat is modified by the current passing through it. Used to test microphones.

thermopile *n.*   A number of thermocouples connected in series.

thermoplastic recording   A process in which information is stored on plastic tape by bombarding it with a modulated electron beam and heating it sufficiently to produce a deformation proportional to the deposited charge. Playback is by means of an optical system. The storage density of the process is very high.

theta (written Θ, θ)   The eighth letter of the Greek alphabet, used symbolically to represent any of various constants, coefficients, etc., of which in electronics the principal ones are: **1** (Θ) Debye temperature. **2** (θ) Temperature. **3** (θ) Angle.

Thevenin's theorem   A theorem stating that with respect to its output terminals any active network can be represented as the series combination of an ideal voltage source $v(t)$ and an impedance Z, where $v(t)$ is the open circuit voltage at the output terminals, and Z is the impedance between these terminals with all independent sources within the network set to zero.

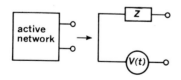

thin film   A film one to several molecules thick.

thin film storage   A form of digital computer storage in which the magnetic elements are thin films deposited on a suitable substrate.

Thomson effect   In a THERMOCOUPLE, the flow of heat when a current is passed across the junction of the dissimilar metals.

thoriated *adj.*   Indicating a tungsten thermionic electrode whose electron emission is enhanced by its being coated thinly with thorium.

thorium (Th)   Element 90; Atomic wt. 232.12; Melting pt. 1845°C; Boiling pt. above 3000°C; Spec. grav. (20°C) 11.3; Valence 4; Spec. heat 0.046 cal/(g)(°C); Electrical conductivity (0–20°C) 0.055×10^6 mho/cm. Radioactive with a half-life of 1.39×10^{10} years, emits alpha particles and yields lead-208. Used in some electronic tubes, and special welding electrodes. Its oxide is used in the tungsten that is employed for electric light filaments.

three-halves power equation   See CHILD-LANGMUIR-SCHOTTLY LAW.

three-phase circuit   A combination of circuits energized by alternating voltages which differ in phase by a nominal 120 degrees, that is, by one third of a cycle.

threshold *n.*   LIMEN

throat *n.*   The narrow part of a HORN.

thulium (Tm)   Element 69; Atomic wt. 169.4; Melting pt. 1545°C; Boiling pt. 1727°C; Density 9.33 g/ml; Valence 3; Spec. heat 0.038 cal/(g)(°C); Electrical conductivity (0–20°C) 0.011×10^6 mho/cm.

thump *n.*   A low-frequency transient, particularly in an audio system.

thyratron *n.*   A gas-filled triode in which a sufficiently large positive pulse applied to the control grid ionizes the gas and initiates conduction. After this time the grid has no further effect, conduction being halted by reducing the plate voltage to zero or less, or by reducing plate current to a value too small to maintain ionization.

thyristor *n.*   A two-, three-, or four-terminal pnpn semiconductor switching device whose bistable characteristics depend on regenerative feedback.

thyrite *n.*   A mixture of silicon carbide with graphite and a ceramic binder, having nonlinear resistance characteristic, used in VARISTORS.

tickler *n.*   In an oscillator circuit, a coil loosely coupled to the inductor of the tuned circuit connected to the output of the active element; its function is to provide the positive FEEDBACK necessary for oscillation.

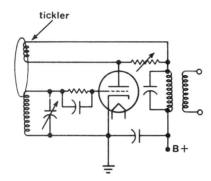

tickler

tie down point A frequency at which the ALIGNMENT of a radio receiver is made.

tie point In circuit wiring, an insulated point where two or more leads are connected.

tight coupling A degree of COUPLING greater than CRITICAL COUPLING.

tilt *n.* **1** The inclination of the horizontal axis of an antenna with respect to the horizon. **2** DROOP.

timbre *n.* The subjective attribute of a sound that correlates most closely with its spectral characteristics.

time base A known and precisely controlled function of time by which some process or phenomenon is controlled or measured, in particular, a linear sawtooth wave used to provide the horizontal deflection of an oscilloscope trace.

time constant The time τ required for a dynamic variable of a system to rise to $1 - 1/e$ (about 63 percent) of its steady-state value or to fall to $1/e$ (about 37 percent) of its steady-state value after a step-function change in excitation. In particular, for RC and LC circuits,

$$\tau = RC, \quad \tau = LC,$$

respectively.

time delay The time required for a signal or wave to travel between two points.

time division multiplex A method of sending several messages through a single communications channel by sampling each message at intervals.

tin (Sn., *Lat.* stannum) Element 50; Atomic wt. 118.70; Melting pt. 231.89°C; Boiling pt. 2260°C;

Spec. grav. (20°) (gray) 5.75, (rhombic 6.55) (tetragonal) 7.31; Hardness (white) 1.5–1.8; Valance 2 or 4; Transformation temp. 13.2°C; Latent heat of fusion 14.2 cal/g; Latent heat of vaporization 520 ± 20 cal/g; Spec. heat (25°C white) 0.053, (10°C gray) 0.049; Heat of transformation 4.2 cal/g; Thermal conductivity (0°C white) 0.150 cal/(cm)(cm²)(°C)(sec); Coefficient of linear expansion (0°C) 19.9×10^{-6}; Shrinkage on solidification 2.8%; Resistivity (0°C white) 11.0×10^{-6} ohm/cm³, (100°C white) 15.5×10^{-6} ohm/cm³; Brinell hardness (20°C) 3.9 10 kg/(5 mm)(180 sec), (220°C) 0.7 10 kg/(5 mm) (180 sec); Tensile strength (15°C as cast) 2100 psi, (200°C) 650 psi, (−40°C) 2900 psi, (−120°C) 12,700 psi. Used as electrical lugs and connectors, and as a constituent of solder. Used also as a protective cladding for other metals.

tipoff *n.* The point at which a vacuum tube is sealed off after evacuation.

titanium (Ti) Element 22; Atomic wt. 47.90; Melting pt. 1800°C; Boiling pt. above 3000°C; Spec. grav. (20°C) 4.5; Valence 2, 3 or 4; Spec. heat 0.126 cal/(g)°C); Electrical conductivity (0–20°C) 0.024×10^6 mho/cm. Used in dielectric materials of capacitors, as some of its compounds have dielectric constants on the order of 10,000. Barium titanate is strongly piezo-electric.

t-m *abbr.* Time modulation.

TM *abbr.* Transverse magnetic. See MODE.

toggle switch A switch in the form of a projecting lever whose movement through a small arc opens or closes an electric circuit.

tone burst A sinusoidal audio signal that is amplitude-modulated with a square wave, used in testing the transient response of audio components.

tone control A variable network used to emphasize or deemphasize a part of the passband of an audio system, thus allowing compensation for deficiencies in program material, room response, etc.

Tophet C An alloy of 24% iron, 15% chromium, 61% nickel. High resistance alloy; oxidation-resistant to about 1010°C.

toroid *n.* A surface having essentially the shape of a doughnut with a hole, generated by revolving a plane, closed curve about a line that lies in its plane but has no point in common with it.

torque *n.* The tendency of a force to produce rotational motion, or, more precisely, the moment **L** of a force **F** with respect to a particular origin, given in vector notation by

$$\mathbf{L} = \mathbf{F} \times \mathbf{r},$$

where **r** is the displacement from the origin of the point at which the force acts.

totem pole amplifier A type of push-pull amplifier that provides a single-ended output signal without the use of a transformer, often used in the output stages of transistor audio amplifiers.

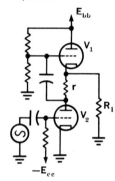

Vacuum tube totem pole amplifier

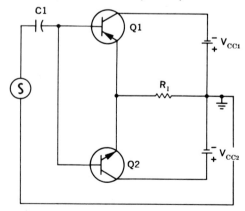

Transistor totem pole amplifier using emitter followers in complementary symmetry.

touch control A circuit that closes a relay when two metal plates are bridged.

tourmaline *n.* A complex cyclosilicate gemstone with both piezoelectric and pyroelectric qualities.

Townsend avalanche An AVALANCHE.

Townsend discharge In a discharge tube, the part of the voltage-current characteristics where the current is low and the applied field alone is not sufficient to maintain the discharge.

TPTG *abbr.* Tuned plated, tuned grid.

TR *abbr.* Transmit-receive.

trace *n.* **1** LINE (def 2). **2** A computer routine which analyzes each operation performed in carrying out a particular instruction.

tracer *n.* A radioactive isotope introduced into a substance in order to determine the subsequent distribution of the substance.

tracing distortion In a disk recording, distortion caused by differences in geometry between the cutting and playback styli.

track **1** *n.* A path, as of the cutting stylus of a phonograph record, a radar target, etc. **2** *v.* To follow the grooves of a phonograph record faithfully. **3** *v.* To maintain a mutually correct rate of variation, as a set of ganged controls, a pair of matched amplifying elements, etc.

train *n.* A sequence, as of waves or pulses.

trans *abbr.* Transmitter.

transadmittance *n.* TRANSFER ADMITTANCE, often as measured under a specified set of conditions.

transceiver *n.* A unit that combines the functions of a radio transmitter and a radio receiver, generally having some of its components common to both operating modes.

transconductance *n.* A transfer function g_m often used to relate the grid (gate) voltage of a vacuum tube (field-effect transistor) to the plate (drain) current. If each of these devices is represented as the same four-terminal network g_m is given by

$$g_m = \partial i_2 / \partial V_1,$$

or

$$g_m \cong \Delta i_2 / \Delta V_1.$$

transconductance tester   A testing device in which the merit of a vacuum tube is evaluated in terms of its transconductance.

transcribe *v.*   In a computer, to transfer information from one external storage to another.

transcription *n.*   The recording of a broadcast.

transducer *n.*   A device by which energy may be coupled from one system into another, in particular, a device of this kind in which the input and output energy are of different types, an example being a loudspeaker, in which electrical energy is converted to audible acoustic energy.

transfer *v.*   **1** In a computer, to transfer data from one storage section to another. **2** To change the location of the next instruction in a computer program.

transfer admittance   An admittance that relates a voltage at one pair of terminals to the current response at another pair.

transfer characteristic   A TRANSFER FUNCTION of an electron device, as a tube or transistor, usually presented in graphical form.

transfer constant   For a network, transducer, etc., terminated in its IMAGE IMPEDANCE, a constant T given by

$$T = \tfrac{1}{2} \ln (E_{in} \, I_{in} / E_{out} \, I_{out}),$$

where **E** is voltage (or an analog thereof) and **I** is current (or an analog thereof) and **E** and **I** may both be complex numbers.

transfer function   In a network, a function that relates the response that results at one pair of terminals to a source applied at another pair of terminals.

transfer impedance   An impedance that relates the current at one pair of terminals to the voltage response at another pair.

transfer ratio   A GAIN FUNCTION.

transfer standard   A movable physical standard, as for instance, a portable frequency meter.

transfluxor *n.*   A magnetic storage or switching element having three legs for flux and several windings.

transform *n.*   In mathematics, an operator that performs the mapping

$$f(t) \longrightarrow g(s).$$

Principally, such operators are integrals of the form

$$g(s) = \int K(t,s)f(t)dt.$$

In electronics, the FOURIER TRANSFORM and the LAPLACE TRANSFORM are the ones most often employed, generally as aids in the solution of equations.

transformer *n.*   A device for transferring ac current from one circuit to another by means of magnetic induction. Each circuit is represented by a winding around a magnetic core, the ratio of the voltages in the two circuits being (in an ideal transformer) equal to the ratio of the number of turns. From Ohm's law and the law of conservation of energy it follows that (except for losses) the product of current and voltage is the same in both circuits.

transient *n.*   Any nonrepetitive waveform.

transient response   The response of a system to a discontinuous change in its input.

transistor *n.*   A three-terminal active semiconductor device consisting of a 'sandwich' in which the outer regions (the COLLECTOR and EMITTER) are of semiconductor material of one type and the enclosed region (the BASE) is of the other type. A transistor can operate as a switch, as in digital systems applications, or as a linear device, as in analog circuit applications. As a switch the transistor alternates between a very low impedance state (saturation) and a very high impedance state (cutoff). In saturation both the emitter-base and collector-base junctions are forward biased, which results in all three terminals being at about the same potential. In cutoff both junctions are reverse biased. Operating as a linear device a transistor is said to be in the active region. In this region the emitter-base junction is forward biased while the collector-base junction is reversed. Under the influence of the forward bias between emitter and base, carriers of opposite polarities are injected into each of these regions, migrate to the junction between them, and undergo RECOMBINATION, the charge displaced due to recombined carriers constituting the base current. Since the rate of carrier injection considerably exceeds the rate of recombination in the base region, most of the carriers migrating from the emitter into the base escape recombination and are attracted by the electric field of the collector region. The carriers injected into the base are repelled by the collector field and can enter the emitter only through recombination; thus the base current is small while the emitter and collector currents are many times larger. Since the rate of carrier injection at the emitter depends on the magnitude of the forward bias between the base and emitter, the base current effectively controls the collector current and amplification occurs. In a moderately high gain transistor something over 90 percent of the emitter current reaches the collector. See N-TYPE SEMICONDUCTOR; P-TYPE SEMICONDUCTOR; PN JUNCTION.

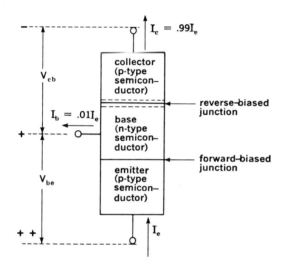

Diagram of a pnp transistor
with $\alpha = .99$

transistor parameters The performance characteristics of a transistor or class of transistors.

transistor seconds In a production run of transistors, those which remain after all those meeting a more stringent set of performance standards have been selected.

transition n. A change, often discontinuous, in the parameters of a wave medium, a conductive medium, etc.

transition frequency In the frequency response curve of a disk recording, the point at which the ASYMPTOTES to the constant-amplitude and constant-velocity portions of the curve intersect. From this point downward bass response is rolled off gradually in order to prevent overmodulation of the grooves.

transition region 1 The condition of a transistor when it is neither cut off nor saturated. 2 DEPLETION REGION.

transitron n. An oscillator circuit employing a tetrode or tetrode-connected pentode operated in its negative resistance region, that is, with its screen grid at a more positive potential than its plate.

translation n. In a computer, the conversion of specific information from one form or language into another; also the reading out of a message.

translator n. A translating routine or device.

transmission n. The conveyance from one point to another of electric power or an electric signal.

transmission line A system of conductors used for the transmission of electric power or signals between two points.

transmission primaries In color television, the set of three independent signals that specify the picture, generally the luminance signal, orange-cyan signal, and the green-magenta signal.

transmit-receive switch In radio and radar, a tube that cuts off the receiver while the transmitter is activated.

transmitter n. The element of a communications or radar system that generates signals, in particular, the part of a radio or television system that generates and modulates the carrier wave.

transponder n. A device that emits radar or radio signals in response to an interrogating pulse.

transrectification n. The production of a direct current or voltage in one circuit as a result of the application of an alternating current or voltage in another. The transrectification factor T is given by

$$T = \partial d/\partial a,$$

where d is the direct current or voltage and a is the alternating current or voltage.

transresistance n. The reciprocal of TRANSCONDUCTANCE.

transverse electric mode See MODE.

transverse electromagnetic mode See MODE.

transverse magnetic mode See MODE.

transverse wave   A wave whose displacement is normal to its direction of propagation. An example is a TEM wave where both the electric and magnetic field are thus oriented.

trap *n.*   **1** A FILTER or resonant circuit used to supress an unwanted signal. **2** In a digital computer, an interruption in the execution of a program indicating an attempt to perform an impossible mathematical operation such as division by zero. **3** An impurity, usually in a silicon semiconductor, that immobilizes a carrier for an unduly long time.

trapezoidal wave   An oscilloscope display in which an amplitude-modulated carrier provides the vertical deflection and the modulating wave the horizontal deflection. The resulting waveform, trapezoidal in shape, can be used to measure the percentage of modulation.

treble *n.*   The upper range of audible frequencies.

tree *n.*   A connected network that for the particular number of nodes that it contains has the minimum possible number of branches.

trf *abbr.*   TUNED RADIO FREQUENCY.

triac *n.*   A bidirectional triode thyristor whose voltage current characteristics are the equivalent of two silicon controlled rectifiers connected in inverse parallel.

triboelectricity *n.*   Electrostatic charges generated by friction.

trichromatic coefficient   One of the coordinates of a particular color on the CHROMATICITY DIAGRAM.

trickle charge   A continuous small charge replenishing a storage battery.

trigatron *n.*   A gas-filled tube, ordinarily non-conducting, in which conduction is initiated by a pulse applied to an auxiliary electrode.

modulation
less than 106%   modulation 100%   overmodulation

traveling wave   A wave propagating through a medium; a wave whose amplitude is a function of time and position. See WAVE EQUATION.

traveling-wave oscilloscope   A special type of cathode ray oscilloscope that achieves a bandwidth of up to 5 ghz by means of a traveling-wave deflection system.

traveling-wave tube   A microwave tube whose operation depends on the interaction of the field of an electromagnetic wave traveling through a waveguide with a beam of electrons that travel with it. If the tube is adjusted so that the phase velocity of the wave is slightly slower than the average velocity of the electrons, the beam is on the average, slowed by the field of the wave, losing energy in the process. It follows from the law of conservation of energy that if the waveguide is not excessively lossy, the energy lost by the beam is gained by the wave.

trigger *v.*   To initiate conduction in (a circuit, device, etc.) by means of a pulse.

trigistor *n.*   A three-terminal pnpn semiconductor device that turns successively on and off in response to pulses of the same polarity applied to its gate.

trigonometric functions   Certain functions of an angle or an arc, of which the most commonly used are the sine, cosine, tangent, cotangent, secant and cosecant. The following statements may be derived from definitions of the functions (see diagram) and the Pythagorean theorem:

$$\tan \theta = \frac{\sin \theta}{\cos \theta}, \ \sin \theta = \frac{1}{\csc \theta}, \ \cos \theta = \frac{1}{\sec \theta}, \ \tan \theta = \frac{1}{\operatorname{ctn} \theta},$$
$$\sin^2 \theta + \cos^2 \theta = 1, \ \tan^2 \theta + 1 = \sec^2 \theta, \ \operatorname{ctn}^2 \theta + 1 = \csc^2 \theta.$$

It can also be shown by other means that

$$e^{i\theta} = \cos\theta + i\sin\theta,\ \cos\theta = \frac{e^{i\theta} + e^{-i\theta}}{2},$$

$$\sin\theta = \frac{e^{i\theta} - e^{-i\theta}}{2i},\ \sin-\theta = -\sin\theta,\ \cos-\theta = \cos\theta.$$

See HYPERBOLIC FUNCTIONS.

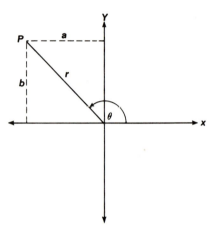

TRIGONOMETRIC FUNCTIONS

If θ is the angle formed by r and the x axis, and P is a point on r having a as its abscissa and b as its ordinate, then sine $\theta = b/r$, cosine $\theta = a/r$, tangent $\theta = b/a$, cosecant $\theta = r/b$, secant $\theta = r/a$, and cotangent $\theta = a/b$.

trim v.   To make a fine adjustment of (a circuit or circuit element).

trimmer n.   A small adjustable circuit element connected in conjunction with a circuit element of the same kind in such a way that its adjustment brings the combination of the two to a desired value.

triode n.   A three-terminal active device, in particular, a vacuum tube having a cathode, plate, and control grid.

tri-tet oscillator   A crystal-controlled OSCILLATOR, in which loading on the crystal circuit is minimized by using the screen grid as the anode of the oscillator circuit and taking the output from the plate.

trochotron n.   BEAM SWITCHING TUBE.

trombone n.   A device for adjusting a coaxial line assembly, shaped somewhat like the musical instrument.

truncate v.   To drop significant digits of (a number), as for instance, to represent 19.7857 as 19.785. Compare ROUND.

trunk n.   A circuit connecting two telephone exchanges.

T.U. abbr.   Transmission unit.

tube n.   An ELECTRON TUBE.

tune v.   To adjust a circuit so that its IMMITANCE is a maximum at some desired frequency or band of frequencies.

tuned amplifier stage   An amplifier stage which by means of resonant circuits included in its input and/ or output produces amplification substantially at a single frequency.

tuned circuit   A circuit having an IMMITANCE function that shows a relative maximum or minimum at some frequency, in particular, the series or parallel combination of an inductance and a capacitance.

tuned radio-frequency   Denoting a type of radio receiver in which amplification is carried out at the carrier frequency rather than at an intermediate frequency as in a SUPERHETERODYNE. Tuning of the receiver is accomplished by varying the parameters of the resonant circuits in each amplifier stage.

tungsten (W; Ger. wolfram)   Element 74; Atomic wt. 183.92; Melting pt. 3410°C; Boiling pt. 5900°C; Spec. grav. (20°C) 19.3; Density (metal powder) 2.0–4.8 g/cm³, (1200°C sintered) 10.0–12.0 g/cm³, (3000°C sintered) 17.0–18.5 g/cm³, (swaged rod) 17.0–19.2 g/cm³, (drawn wire) 19.3 g/cm³; Valence 2, 4, 5, or 6; Latent heat of fusion 45 cal/g; Latent heat of vaporization 1180 cal/g; Spec. heat (20°C) 0.032 cal/(g) (°C); Thermal conductivity (0°C) 0.40 cal/(cm)(sec) (°C), (1227°C) 0.28 cal/(cm)(sec) (1827°C) 0.40 cal/ (cm)(sec)(°C); (2227°C) 0.50 cal/(cm)(sec)(°C); Linear thermal expansion $L = L_0 [1 + (4.28t + 0.00058t^2) \times 10^{-6}]$, $t = °C$; Vapor pressure (1727°C) 3×10^{-15} atm, (3410°C) 5.4×10^{-6} atm, (4227°C) 5.6×10^{-3} atm; (5727°C) 0.9 atm; Electrical resistivity (20°C) 5.6×10^{-6} ohm/cm, (927°C) 30.2×10^{-6} ohm/cm, (1827°C) 59.0×10^{-6} ohm/cm, (2727°C) 90.4×10^{-6} ohm/cm. Pure tungsten metal is used as wire, rod and sheet in the electric lamp, electronics and electrical industries; Used for welding rods, lead wires, cathodes for power tubes, radio tubes, electric wiring for furnaces, and almost exclusively in light filaments. Silver or copper-tungsten alloys are used for heavy-duty electrical contacts.

tungsten steel   An alloy of iron, carbon (0.7%), manganese (0.3%) and tungsten (5%) used for permanent magnets, having a coercive force of 70 oersteds and a remanence of 10,300 gauss.

tunnel diode   A two-terminal semiconductor device containing a single, abrupt pn junction made with DEGENERATE SEMICONDUCTOR MATERIAL and having, as a result, in the region of low forward bias a NEGATIVE RESISTANCE characteristic.

tunnel effect   A quantum mechanical effect by which a particle has a finite probability of penetrating a POTENTIAL BARRIER. The magnitude of the probability increases with the energy of the particle and decreases with the thickness of the barrier.

turnoff delay time   STORAGE TIME.

turnoff time   In a switching circuit, the time taken to cut off the current after a turnoff signal.

turnon delay time   The time taken by a device to reach 10 percent output after receiving an input pulse.

turnover frequency   TRANSITION FREQUENCY.

turnstile antenna   An antenna composed of two crossed dipoles normal to each other fed with currents that are in phase quadrature.

turntable n.   The rotating disk that carries a phonograph record.

turret tuner   In a television receiver, a tuner having a separate resonant circuit for each channel.

TVI abbr.   Television interference.

tweeter n.   A loudspeaker designed especially for the reproduction of high-frequency sounds.

twin lead   TWIN LINE.

twin line   A form of transmission line consisting of two parallel conductors imbedded in a dielectric material, the characteristic impedance of the line being determined by the diameter and spacing of the conductors and the dielectric constant of the surrounding material.

twinning n.   **1** A process in which one region of a crystal orients itself in a way that has a symmetrical relation to the rest of the crystal. **2** In television, a fault condition in which the interlaced alternate fields are incorrectly spaced.

two-phase circuit   A pair of circuits energized by alternating voltages whose phases differ by 90°

TWS abbr.   Timed wire service.

TWT abbr.   Traveling-wave tube.

U

UHF, uhf abbr.   ULTRA-HIGH FREQUENCY.

ultor n.   In a television cathode-ray tube, the predeflection acceleration electrode that is set at the highest positive potential.

ultra-high frequency   A radio frequency in the band between 300 and 3,000 Mhz.

ultralinear amplifier   A CLASS AB or B audio power amplifier using pentodes or beam power tubes whose screen voltages are taken from taps on the output transformer rather than from a fixed dc source. This form of operation results in a considerable decrease in distortion.

ultrashort wave   A radio wave between 1 and 30 meters in length.

ultrasonic adj.   Indicating a sound wave having a frequency in excess of about 20 khz.

ultrasonics n.   The branch of physics and technology dealing with acoustic phenomena above the audible frequency range.

umbrella antenna   An antenna whose elements point downward like the spokes of an umbrella.

uncertainty principle   **1** In quantum mechanics, a principle stating that the position **r** of a particle and its momentum **p** are related in such a way that

$$\Delta\mathbf{p} \cdot \Delta\mathbf{r} \cong h,$$

where $\Delta\mathbf{p}$, $\Delta\mathbf{r}$ are respectively the minimum errors in the respective measurements of **p**, **r** and h is Planck's constant. The principle also postulates the same relation between energy and time. **2** In information theory, a similar relation that arises between bandwidth and time (*T*). Defining bandwidth as ΔF, where *F* is frequency, the relation states that

$$\Delta F \cdot \Delta T \cong 1.$$

undercut *n.*   Removal of some of the conductive metal in a printed circuit as the result of overexposure to the etchant.

underflow *n.*   In a digital computer, the generation of a negative number whose magnitude exceeds the capacity of a register, storage location, etc.

underlap *n.*   A gap in a FACSIMILE system.

undershoot *n.*   The initial response of a device to a change in input, opposite in direction to and preceding the main transition.

uniconductor *n.*   A WAVEGUIDE with a rectangular or cylindrical conductive surface surrounding a homogeneous dielectric.

unidirectional *adj.*   Of an antenna, having maximum GAIN in a single direction.

unijunction transistor   A three-terminal semiconductor device consisting of a bar of n-type silicon having a resistance gradient between its ends (which are ohmic contacts called *base-one* and *base-two*) with a rectifying contact (called the emitter) between base-one and base-two. In operation, with base-one grounded and positive voltage applied to base-two, the emitter junction remains reverse-biased so long as the voltage applied there is less than that produced by the voltage divider action due to the current flow through the silicon bar. When the emitter is forward-biased holes are injected into the silicon bar with the simultaneous generation of mobile electrons equal in number to the holes. As a result of this, the resistance between the emitter and base-one decreases, and the emitter to base-one voltage falls. Thus there is a negative resistance region in the voltage-current characteristic between the emitter and base-one.

unilateralization *n.*   The operation of a vacuum tube or transistor in conjunction with a network that cancels both the real and imaginary parts of its internal positive feedback.

unipolar transistor   A transistor, as for example a FIELD-EFFECT TRANSISTOR, in which all the current carriers are of the same polarity.

unit pole   A MAGNETIC POLE of such strength that, if placed one centimeter away from a pole of like strength, the force between them will equal one dyne.

unit pulse   A BAUD.

unitunnel diode   A TUNNEL DIODE with peak reverse currents in the microampere region.

unity coupling   **1** Perfect magnetic coupling between two coils, that is, coupling in which there is no leakage flux. **2** An interaction between two systems in which all the energy in one is transferred to the other, as, for example, in an ideal transformer.

universal motor   An electric motor that will operate with ac or dc current at essentially the same speeds and torques.

universal receiver   A radio receiver that will operate on either dc or ac current.

univibrator *n.*   MONOSTABLE MULTIVIBRATOR.

unstable equilibrium   See EQUILIBRIUM.

uranium (U)   Element 92; Atomic wt. 238.07; Melting pt. about 1150°C; Boiling pt. 3818°C; Spec. grav. (20°C) 18.68; Valence 3, 4, or 6; Heat of fusion 407 kcal/mole; Heat of vaporization 106.7 kcal/mole; Spec. heat 0.028 cal/(g)(°C); Heat capacity (25°C) 6.612 cal/(°C)(mole); Thermal conductivity (70°C) 0.071 cal/(cm-sec)(°C); Electrical conductivity (0–20°C) $2-4 \times 10^4$ mho/cm. Radioactive element. Used as a nuclear fuel.

uv *abbr.*   Ultraviolet.

V

V *symbol*   **1** (**v**) Velocity. **2** (*V*, *v*) Volume. **3** (*V*) Potential. **4** (*V*) Verdet constant. **5** (*V*, *v*) Volt(s). **6** (*v*) Voltage.

VA *abbr.*   Volt-ampere(s).

vacuum tube   An ELECTRON TUBE with a high internal vacuum.

vacuum-tube voltmeter   A voltmeter in which the measured voltage is amplified in a vacuum tube circuit before being applied to the meter. One of the principal advantages of the instrument is its high input impedance. Many meters of this kind have provision for connection as ohmmeters.

valence *n.* The property of an atom which determines with what and how many other atoms it may combine, based on the numbers of electrons in its outer shell.

valence band The highest electronic energy band in an unexcited solid, separated, in the case of insulators and semiconductors, from the conduction band by a band of forbidden energy levels. See BAND THEORY OF SOLIDS.

valley *n.* A dip in a curve.

valley point In a tunnel diode, unijunction transistor, or similar negative resistance device the value of voltage at which the current has a minimum value.

valve *n.* *British* Electron tube.

vanadium (V) Element 23; Atomic wt. 50.95; Melting pt. 1710°C; Boiling pt. 3000°C; Spec. grav. (20°C) 5.96; Valence 2, 3, 4, or 5; Spec. heat 0.120 cal/(g)(°C); Electrical conductivity (0–20°C) 0.04 $\times$ 10^6 mho/cm. Used as a constituent of many steels.

Vanadium Permendur An soft magnetic alloy of 49.2% iron, 1.8% vanadium, and 49% cobalt, having qualities similar to PERMENDUR.

Van der Pol equation An equation, namely

$$x'' - \varepsilon(1 - x^2)x' + x = 0,$$

that describes the main features of oscillations in non-conservative systems, giving, in fact, a good approximation of the behavior of NEGATIVE RESISTANCE oscillators. For values of the parameter $\varepsilon \ll 1$ the periodic solutions are of the quasi-harmonic type; for $\varepsilon \gg 1$ relaxation-type solutions occur.

V-antenna A directional antenna shaped in the form of the letter V.

var *n.* A reactive volt-ampere.

varactor *n.* A semiconductor diode whose junction capacitance is its principal parameter of interest. Since this capacitance decreases as the reverse bias across the junction is increased, a device of this kind represents a convenient voltage-controlled capacitor having applications in frequency modulators, parametric amplifiers, etc.

variable *n.* A quantity to which any one of a given set of values may be assigned.

variable-μ tube A REMOTE CUTOFF TUBE.

Variac A variable autotransformer: A trade name.

varicap *n.* A VARACTOR.

varindor *n.* An inductor whose value undergoes large variations as the current through it changes.

variocoupler *n.* A radio-frequency transformer with an adjustable coil allowing changes in the coupling.

variolosser *n.* A device whose LOSS can be to some extent controlled by a voltage or current.

variometer *n.* A variable inductor consisting of a pair of series- or parallel-connected coils whose axes can be varied one with respect to the other, the change in mutual inductance causing a change in the total inductance.

varistor *n.* Any of various semiconductor devices whose resistance is a function of impressed voltage, in particular, a symmetrical device having this characteristic.

Varley loop A variation of the MURRAY LOOP using a variable resistance.

V-cut crystal A slab of quartz in which no cut is made parallel to an axis of the original crystal.

Vectolite *n.* A FERRITE containing 30% Fe_2O_3, 44% Fe_3O_4, and 26% CoO_3, with a higher coercive force than any material except alnico XII.

vector *n.* A mathematical quantity, as displacement, velocity, acceleration, etc., that is uniquely specified by its magnitude and direction. A more careful definition of a vector requires that it be invariant under a translation or rotation of the coordinate system in which it is described. For example, assume a quantity **V** whose components are x_1, x_2, x_3. If the coordinate system is rotated while the origin remains fixed the new components are given by

$$x_i' = \sum_{j=1}^{3} c_{ij} x_j, i = 1,2,3,$$

where c_{ij} are the direction cosines between the coordinate axis pairs. **V** is a vector if and only if this transformation applies. See PSEUDOVECTOR.

vector diagram The representation of an alternating quantity as a PHASOR.

vector product The vector **V** (more precisely a PSEUDO-VECTOR) indicated by

$$V = A \times B$$

where **A, B** are vectors. Its magnitude is given by

$$|V| = |A| \, |B| \sin \theta$$

where θ is the angle between **A** and **B**. Its direction is defined in the following way—let a right-handed coordinate system be rotated so that **A, B** lie in the xy-plane and let the positive sense of θ be taken as counter-clockwise, then the positive sense of **V** coincides with the positive sense of the z-axis.

velocity modulation The variation of the velocity of a beam of electrons by an electrostatic field that is periodic with respect to time and inhomogeneous with respect to space, with the result that the electrons become bunched as they travel from cathode to anode. The operation of KLYSTRONS and TRAVELING-WAVE TUBES depends on this effect.

vernier *n*. An auxiliary adjustment or measuring device providing finer readings than the large scale measure to which it is attached.

very low frequency A radio frequency in the band between 10 and 30 khz.

vestigial sideband An amplitude-modulation system in which one sideband is reduced but not compressed. It allows lower frequencies to be transmitted than in single sideband transmission. The carrier may or may not be suppressed.

VFO *abbr*. Variable frequency oscillator.

VHF, V.H.F. vhf *abbr*. Very high frequency.

vibrating reed frequency meter A frequency meter consisting of a set of tuned tempered steel reeds which are vibrated electromagnetically in step with the supply voltage, that reed which vibrates most vigorously being in resonance with the supply voltage.

vibrator *n*. A mechanical CHOPPER for direct current.

video *n*. The visual portion of a television broadcast, signal, etc.

video amplifier An amplifier having a rather broad passband, generally from about 15 hz to about 5 Mhz.

video tape Magnetic tape on which a television program may be recorded.

Villari effect The change in magnetic induction in an iron rod caused by longitudinal stress. See MAGNETO-STRICTION.

virtual ground In an operational amplifier, the virtually zero input impedance produced by heavy FEEDBACK. Although a voltage approaching zero appears across the input terminals the current drawn is negligible.

virtual image In geometrical optics, the point at which rays from a point source would meet if, after a divergent refraction, their new courses were projected backwards.

visible speech 1 The representation of speech in visible form by the use of sound spectrograms. **2** Phonetic symbols designed to represent every possible position taken by the speech mechanism in the process of articulation.

VLF, vlf *abbr*. Very low frequency.

vocoder *n*. An instrument for encoding speech into electrical signals which can be reproduced into approximately equivalent sounds after transmission. The device reduces the bandwidth required for a voice communications channel by a factor of approximately 10, but in the process makes the speaker unidentifiable.

vodas Voice-operated device, anti-sing: a device in a telephone circuit which prevents SINGING by allowing transmission in only one direction at a time.

voder *n*. A device for producing synthetic speech, consisting essentially of an array of key-controlled oscillators and filters.

vogad Voice-operated gain-adjuster device: a device attached to the input of a sound system regulating the gain so as to produce a relatively constant level of output.

voice coil The moving coil used to drive the cone of a loudspeaker.

Voiceprint *n*. A type of speech spectrograph that is sufficiently sensitive to identify individual human voices.

volt *n*. The measure of electromotive force, defined as the difference of potential which, when applied across a resistance of 1 ohm, produces a current of 1 ampere.

voltage *n.* The difference in electrical POTENTIAL between two points; electromotive force.

voltage divider A four-terminal network in which the output voltage is a proper fraction of the input voltage. In the most usual type the input is applied across two resistors in series, the output being taken across one of them.

voltage doubler A rectifying circuit with a dc output approaching twice that of the peak ac input.

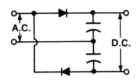

voltage regulator A device for keeping a voltage relatively constant.

volt-ampere *n.* The product of current and voltage in an ac circuit, in a purely resistive circuit, identical with power. In a circuit containing reactance power is given by the product of current, voltage, and the POWER FACTOR.

voltmeter *n.* An instrument for measuring voltage. In order to produce an accurate measurement it is necessary that the input impedance of the instrument be at least 10 times the source impedance at the terminals under test.

volt-ohm-milliammeter A test instrument for measuring dc and ac voltages, resistances, and currents.

volume *n.* **1** The magnitude of a complex electric wave, such as one corresponding to sound. **2** The space taken up by an object.

volume unit A DECIBEL defined with respect to a reference level of 1 milliwatt into 600 ohms as indicated on a standard type of meter having specified ballistic characteristics.

volume velocity The rate of flow of a medium disturbed by a sound wave.

VOM, vom *abbr.* Volt-ohm-milliammeter.

voting gate MAJORITY GATE.

VR tube A VOLTAGE REGULATOR tube.

VSB *abbr.* Vestigial sideband.

VSWR *abbr.* Voltage standing-wave ratio.

VT *abbr.* VACUUM TUBE.

VTVM *abbr.* VACUUM-TUBE VOLTMETER.

VU *abbr.* VOLUME UNIT.

W

W *symbol* **1** (w_g) Gross work function. **2** (*w*) Net work function. **3** (*W*) Radiant flux density.

W *abbr.* **1** Watt(s). **2** Work.

wafer socket A vacuum tube socket consisting of a wafer of insulating material with holes in it in which the tube is inserted and its connections gripped by spring clips.

wafer switch A device with one or more insulated disks with connections on their circumference which can be switched by a rotating arm.

Wagner ground In an ac BRIDGE, a special network through which a ground connection is made in order to minimize errors due to stray capacitance.

waiting time In certain electron tubes, the period that elapses between the time when the heater is switched on and the time at which plate voltage can be safely applied.

walkie-lookie *n.* A portable television camera and transmitter.

walkie-talkie *n.* A portable radio transceiver.

wamoscope *n.* A cathode ray oscilloscope equipped to detect and display radar signals.

warble-tone generator An audio signal generator whose frequency is modulated over a small range at a subaudio rate, used to minimize spurious measurements resulting from standing waves.

warm up The time taken by a system to come into full operation after power has been applied.

watt *n.* The measure of power, defined as work being done at the rate of one JOULE per second. Electrically this is equivalent to the power developed when one AMPERE flows through a resistance of one OHM.

watt-hour The measure of energy equivalent to that represented by one watt acting for one hour.

wattmeter *n*. An instrument for measuring electric power.

watt-second JOULE.

wave *n*. A disturbance whose magnitude is a function of time and position. See ELECTROMAGNETIC WAVE; WAVE EQUATION.

wave equation A partial differential equation describing the propagation of a wave, namely

$$\Delta^2 u = \frac{1}{v^2} \frac{\partial^2 u}{\partial t^2}$$

where u is the dynamic variable of the wave, v is its speed of propagation, and t is time.

waveform *n*. The graphical representation of the shape of a wave with amplitude plotted against time.

wavefront *n*. **1** A surface at every point of which a wave is at the same phase at a given time. **2** The portion of the wave ENVELOPE (def. 1) between its virtual zero and its crest.

wavelength *n*. The distance, measured along the line of propagation, between two points representing similar phases of two consecutive waves, given by

$$\lambda = c/f$$

where λ is the wavelength, c the speed of propagation, and f the frequency of the wave.

wavemeter *n*. A device for measuring the wavelength and frequency of an electromagnetic wave by means of resonance effects.

wave soldering A process for connecting components to a printed circuit board. After the components are mounted and the copper side of the board fluxed, the board rides on a conveyor across a tank which is so agitated that the peak of a standing wave of molten solder just makes contact with the copper side of the board.

wave train A series of waves propagating from a source.

Wb *abbr*. Weber(s).

wearout *n*. The point at which a device has reached the end of its useful life and is not worth repairing.

wedge *n*. **1** A tapered length of carbon at the end of a waveguide. **2** A trapezoidal television test pattern. **3** An optical filter in which the transmission decreases from one end to the other.

Wehnelt cathode A thermionic cathode whose electron emission has been enhanced by a coating of a metal oxide of the alkaline earth group.

weighting *n*. The adjustment of values of a measured variable in order to compensate for various effects such as conditions of use that vary from the conditions of measurement, varying uncertainties of measurement, etc.

Wertheim effect The production of voltage across the ends of a ferromagnetic wire when it is twisted in a longitudinal magnetic field.

w.g. *abbr*. Wire gauge.

Wh, wh, whr *abbr*. Watt-hour.

Wheatstone bridge A BRIDGE circuit in which all four branches are resistances, used for the measurement of resistance.

whip antenna A type of antenna consisting of a single, flexible metal pole, used mainly on motor vehicles.

whisker *n*. A thin single crystal filament either grown in a supersaturated solution or spontaneously extruded from a mass of material, having relatively high strength and unusual electric and magnetic properties.

whistler *n*. A transient radio signal in the VERY LOW FREQUENCY range, resulting from the dispersion of an impulse generated by a stroke of lightning.

white light Light that has an even distribution of wavelengths throughout the visible spectrum; light that is approximately equivalent to noon sunshine.

white noise Random noise with equal energy distribution at all frequencies.

whr.m. *abbr*. Watt-hour meter.

Wien bridge An ac BRIDGE in which two adjacent branches are resistances, one branch is a series combination of resistance and capacitance, and the other a parallel combination of resistance and capacitance. Its applications include the measurement of frequency

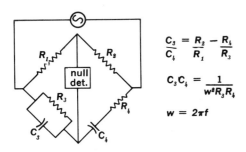

$$\frac{C_3}{C_4} = \frac{R_2}{R_1} - \frac{R_1}{R_3}$$

$$C_3 C_4 = \frac{1}{w^2 R_3 R_4}$$

$$w = 2\pi f$$

Wien bridge for measuring capacitance.

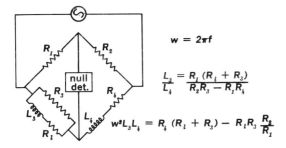

$$w = 2\pi f$$

$$\frac{L_3}{L_4} = \frac{R_1 (R_1 + R_3)}{R_2 R_3 - R_1 R_4}$$

$$w^2 L_3 L_4 = R_4 (R_1 + R_3) - R_1 R_3 \frac{R_2}{R_1}$$

Wien bridge for measuring inductance.

and capacitance, and the determination of the operating frequency in certain oscillator circuits.

Williamson amplifier An audio power amplifier developed by D. I. N. Williamson using tetrodes or beam-power tubes connected as triodes in a push-pull output stage.

wind *n.* The way in which tape is wound on a spool; in an A wind the magnetic surface faces inward, in a B wind, outward.

winding *n.* The manner in which a conducting wire is wound in a coil.

wire gauge A system of classifying wires by diameter. See AMERICAN WIRE GAUGE.

wireless microphone A microphone connected to a small radio transmitter that transmits a signal to a suitable receiver located a short distance away.

wirephoto *n.* **1** A photograph coded into electrical

signals and transmitted by wire. **2** The method by which this is done. See FACSIMILE.

wire recorder A device for recording sounds by electromagnetic registration on a fine moving wire.

wire-wound resistor A resistor consisting of a length of high-resistance wire wound on a supporting form.

w.l. WL, wl, W.L. *abbr.* Wave length.

wobbulator *n.* A signal generator whose output frequency is swept by a motor driven variable reactance device.

woofer *n.* A loudspeaker designed to reproduce the low end of the audio spectrum.

word *n.* In a digital computer, a group of digits or characters that are contained in one storage location.

work *n.* The energy transferred by a force in moving a body through a given displacement, more precisely work W is given by

$$W = \int_a^b \mathbf{F} \cdot \mathbf{dr}$$

where **F** is the force, **r** the path of displacement, and *a,b* the limits of the path.

work function The energy needed to remove an electron from the Fermi level of a metal to an infinite distance outside of its surface. See FERMI-DIRAC STATISTICS.

working voltage VOLTAGE RATING.

wow *n.* In an audio reproduction system, a frequency modulation of the signal, caused, in general by irregular movement of some mechanical part.

wpc *abbr.* Watts per candle.

wrap-around storage In a digital computer, a layout of core storage in which the lowest numbered storage location follows the highest numbered one, that is, a core storage that is layed out in a ring.

writing speed In a cathode-ray tube, the maximum linear speed at which the electron beam can produce a visible trace.

wt *abbr.* Weight.

WWV A radio station operated by the National Bureau of Standards broadcasting, among other services, frequency signals accurate to one part in 5×10^7.

WWVH A radio station of the Bureau of Standards similar to WWV but situated in Hawaii.

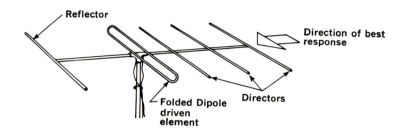

wye *n*. A network consisting of three branches meeting at a common node; an alternate form of TEE network.

X

X *symbol* **1** (*x*) A rectangular coordinate. **2** (*X*) Reactance. **3** (*X*) Acoustical volume displacement.

x-axis **1** In a system of Cartesian coordinates, the axis normal to the y–z plane. **2** In a quartz crystal, a reference axis chosen so as to connect two opposite vertices of its hexagonal cross section; one of the axes showing the greatest electrical activity.

X-cut crystal A slab of quartz cut parallel to the *x*-axis of the original crystal.

xi (written *Ξ, ξ*) The fourteenth letter of the Greek alphabet, used symbolically to represent any of various constants, coefficients, etc., of which in electronics the principal ones are: **1** (*Ξ*) Propagation flux density. **2** (*ξ*) Displacement component of a sound-bearing particle.

xenon (Xe) Element 54; Atomic wt. 50.95; Melting pt. −112°C; Boiling pt. −107.1°C; Spec. grav. (−109°C liquid) 3.52; Density 5.85 g/l; Valence 0. Used to fill flash bulbs in cameras, arc-lamps; readily absorbs radiation.

x-rays *n. pl.* Electromagnetic radiations with a wavelength of 0.1 to 100 angstroms, created by bombarding a metal target with high-velocity electrons.

x-ray tube A vacuum tube used to produce x-rays.

X unit A measure of wavelength (of x-rays), equal to 0.001 Angstrom or 10^{-11} cm.

X-wave The extraordinary wave component of a radio wave that has interacted with the ionosphere.

xy recorder A device that traces on a chart (a Cartesian xy-plane) the relation between two variables.

Y

Y *symbol* **1** (*y*) A rectangular coordinate. **2** (*Y*) Admittance. **3** (*y*) Transverse acoustical displacement. **4** (*Y*) Young's modulus of elasticity.

Yagi-Uda array A directional antenna array consisting of a number of dipoles arranged in a line, one or more of these being DRIVEN ELEMENTS, the others being PARASITIC ELEMENTS. See top of page for diagram.

y-axis **1** In a Cartesian coordinate system, the axis normal to the x–z plane. **2** In a quartz crystal, a reference axis chosen so as to connect the midpoints of two of the opposite sides of its hexagonal cross section; one of the axes on which the crystal's mechanical activity is greatest.

Y-bar A quartz crystal having a Y-CUT.

Y-cut crystal A slab of quartz cut parallel to the *y*-axis of the original crystal.

Y-parameters The parameters of a four-terminal or other device expressed as a set of admittances.

ytterbium (Yb) Element 70; Atomic wt. 173.04; Melting pt. 1490°C; Boiling pt. 2500°C; Spec. grav. 5.51;

Valence (°C); Electrical conductivity (0–20°C) 0.035 × 10⁶ mho/cm.

yttrium (Y) Element 39; Atomic wt. 88.92; Melting pt. 1490°C; Boiling pt. 2500°C; Spec. grav. 5.51; Valence 3; Spec. heat 0.071 cal/(g) (°C); Electrical conductivity (0–20°C) 0.019 × 10⁶ mho/cm.

Z

Z *symbol* **1** (z) A rectangular coordinate. **2** (Z) Impedance.

z-axis **1** In a system of Cartesian coordinates, the axis normal to the x-y plane. **2** In a quartz crystal, the optical axis.

Z-axis modulation In a cathode ray tube, the modulation of electron beam intensity.

Z-bar A quartz crystal having a Z-CUT.

Zeeman effect The splitting of a spectral line from a radiating atom that is subjected to a weak magnetic field into two or more components.

Zener diode A silicon diode designed to undergo abrupt AVALANCHE BREAKDOWN at a predictable reverse bias voltage.

Zener effect The breakdown of a pn junction that is subjected to an electric field strong enough to directly rupture the covalent bonds of the semiconductor material.

Zener voltage The voltage at which the Zener effect takes place.

zero *n.* **1** The identity element under addition, that is, a number (0) having the properties that

$$a + 0 = 0 + a = a$$

and

$$a \times 0 = 0 \times a = 0.$$

2 A value of the argument of a function for which the function takes the value 0. Compare POLE.

zero beat The effect produced when two mixed frequencies are the same and thus have a BEAT FREQUENCY of zero.

zeta *(written* Z, ζ*)* The sixth letter of the Greek alphabet, used in electronics to denote (ζ) a displacement component of a sound-bearing particle.

zinc (Zn) Element 30; Atomic wt. 65.38; Melting pt. 419.5°C; Boiling pt. 907°C; Spec. grav (20°C) 7.14; Hardness 2.5; Valence 2; Spec. heat 0.0915 cal/(g) (°C); Electrical conductivity (0–20°C) 0.167 × 10⁶ mho/cm. Used as plates for primary batteries, and as negative electrodes in certain other types of electric batteries.

zirconium (Zr) Element 40; Atomic wt. 91.22; Melting pt. 1900°C; Boiling pt. above 2900°C; Spec. grav. (20°C) 6.4; Valence 4; Spec. heat 0.066 cal/(g)(°C); Electrical conductivity (0–20°C) 0.024 × 10⁶ mho/cm. Has ability to withstand high temperatures, has low thermal expansivity and good abrasion resistance.

Z-parameters The parameters of a four terminal or other device expressed as a set of impedances.

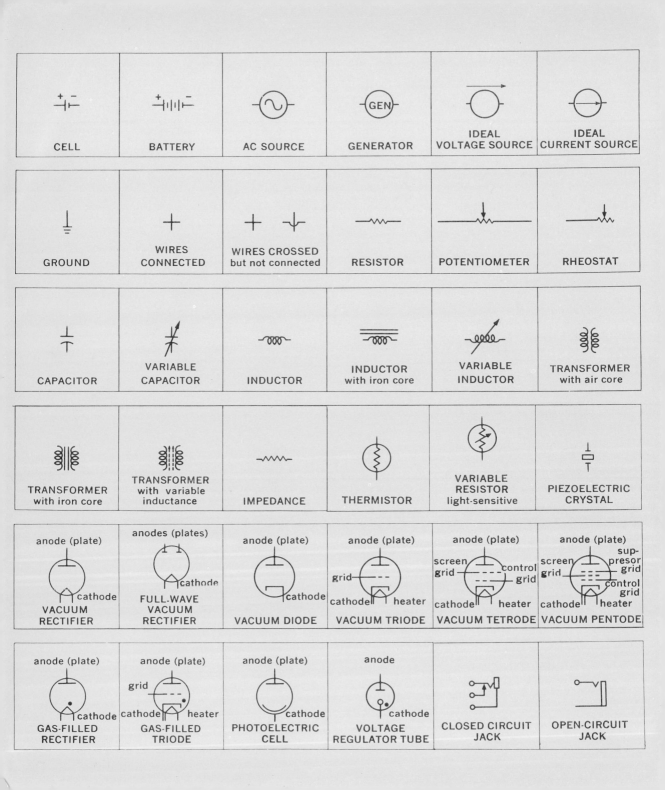